THE
HUMAN SOUL (LOST) IN TRANSITION
AT THE
DAWN OF A NEW ERA

By Erel Shalit

With Contributions by
Nancy Swift Furlotti

H CHIRON PUBLICATIONS • ASHEVILLE, NORTH CAROLINA

The Human Soul (Lost) in Transition At the Dawn of a New Era is published in cooperation with Recollections, LLC. Recollections is devoted to promoting and supporting the publication of material related to the early development of analytical psychology.

www.ChironPublications.com

Interior and cover design by Danijela Mijailovic
Printed primarily in the United States of America.

Cover image by Mira Raman, copyrighted, *The Fifth Day* from the Genesis series, 2007 acrylic on multilayered handmade paper, 32" x 32".

ISBN 978-1-63051-682-6 paperback
ISBN 978-1-63051-683-3 hardcover
ISBN 978-1-63051-684-0 electronic
ISBN 978-1-63051-685-7 limited edition paperback

Library of Congress Cataloging-in-Publication Data

Names: Shalit, Erel, author. | Furlotti, Nancy Swift, contributor.
Title: The human soul (lost) in transition : at the dawn of a new era / by
   Erel Shalit ; with contributions by Nancy Swift Furlotti.
Description: Asheville, N.C. : Chiron Publications, [2019] | Includes
   bibliographical references and index.
Identifiers: LCCN 2018053372| ISBN 9781630516826 (pbk. : alk. paper) | ISBN
   9781630516833 (hardcover : alk. paper)
Subjects: LCSH: Psychoanalysis--Philosophy. | Image (Philosophy) |
   Postmodernism. | Soul.
Classification: LCC BF175 .S486 2019 | DDC 150.19/5--dc23
LC record available at https://lccn.loc.gov/2018053372

Table of Contents

Preface

This book has been in the works for years, evolving, expanding, deepening as technology has taken off through the first decade of our century and continues to soar forward faster than we ever imagined. I encouraged Erel Shalit to finish it, as it was and remains important and timely for people to hear his voice on this subject, to see what he saw, both warnings and ways to adapt. The evidence has become starker, the postmodern experience has changed everyone, along with our perceptions of our lives and future.

Erel's life was unfortunately cut short, contracting cancer in midlife and succumbing to the disease. Was it pollutants in the environment, was it stress, was it happenstance, was it his fate? One cannot answer these questions with any certainly. One can only observe and remain in the dis-ease/dying process with all the emotion and logic and intuition and sensation that bombards the psyche/soul. Not being able to stop the forward rush of mutating, migrating cells, even with the latest technological advances in medicine, he finally came to peace with the inevitable loss of his body reality, his body soul.

His process was an interesting parallel to what he worked so diligently to understand about our evolving world. The ultimate fear regarding AI is losing our bodies, finding them to be redundant. The ultimate question in all this has to do with soul—body soul, spirit soul. It was clear to all who knew him that Erel's spirit and soul remained strong as a result of his conscious efforts to build a compassionate, ethical container for the essence of his being. He was not certain this intentionality would remain an important factor for humanity with the advances he was seeing in the field of technology.

Erel was a compassionate and astute observer of life. He understood the world of instincts as well as the archetypes, and where and when they touch down in reality in the form of our complexes and shadow that we struggle with endlessly. His curiosity and perspicacity was exceptional in exploring how technology fits into our human lives with our million-year-old patterns of behavior and expectations. He was deeply concerned with what he saw in his consulting room and in the world at large.

As his partner, collaborator and best friend, Erel asked me to finish this book for him, which I have done. The topic had been a constant part of discussions over the years and of grave concern to both of us. He pulled all the chapters together and edited a good portion of the manuscript himself but still left quite a bit for me to do. I have to say, immersing myself in his beautiful, poetic writing has been a great distraction from missing him. It is important to note that he included two of my papers in his book. One was a lecture I gave back-to-back with him on a similar topic at the IAAP[1] Congress in Copenhagen in 2013, while another was on my own long-standing interest in Mesoamerican mythology and the environment.

It has been an honor to be able to shepherd this book to its completion for Erel. I know he would be delighted to know that his voice is now loosed upon our world, focusing his crystal-clear intellect and compassionate feeling on a problem that affects us all, a problem that can be a game-changer for humanity either in a creative or destructive way.

Nancy Swift Furlotti, 2018

---

[1] AI refers to Artificial Intelligence. IAAP refers to International Association for Analytical Psychology

Introduction

"I am indeed convinced that creative imagination is ... the real Ground of the psyche, the only immediate reality."

C.G. Jung[2]

He who possesses the world but not its image possesses only half the world, since his soul is poor and has nothing. The wealth of the soul exists in images.

C.G. Jung[3]

The aim of this book is to present a depth psychological perspective on phenomena pertaining to the present, postmodern era. As such, its origins are in the depths; symbolically, in the depth of the waters, in which the sacred is reflected. Likewise, this book centers around the *image*, which has traveled from the forbidden zone of the transcendent command "make no graven image,"[4] through the interiority of the human soul, to become an exteriorized, computerized, robot-generated image that virtualizes as well as augments reality.

---

[2] C.G. Jung, *Letters* Vol. I, p. 60.
[3] Jung, *The Red Book*, (The Readers Edition), pp. 129-130.
[4] Exodus 20:4.

When listening to the brilliant young men and women at the forefront of today's applied science, those who hold the trigger to the chips and the apps of the future, one cannot refrain from being amazed at their confident conviction that technology is the remedy of all ills and the foundation of all future virtues, unaware that unrestrained technology may equally be the platform for future viruses.

I do admit that this book is an attempt to cause a slight crack in the confidence, to whisper in the ears of the triumphant heroes of tomorrow's world, "You are also mortals." I believe that many of them, young, brilliant, and victorious, are unaware that unimpeded promises for the future are inevitably accompanied by no less powerful fears of the apocalypse, fears that, when least expected, rise like monsters from the shadows.

While I am not a Luddite, I cannot help but recall how environmentalists a century ago welcomed the automobile:

> In cities and in towns the noise and clatter of the streets will be reduced, priceless boon to the tired nerves of this overwrought generation. ... On sanitary grounds too the banishing of horses from our city streets will be a blessing. Streets will be cleaner, jams and blockages less likely to occur and accidents less frequent, for the horse is not so manageable as a mechanical vehicle.[5]

We cannot, need not and should likely not return to the days of horse-drawn carriages, but rather welcome much of what technology and scientific progress can offer. Yet, we should perhaps help the architects of the future to balance their sometimes lofty towers of Babel. While they bring us apps for every aspect of life, the elders of society may provide insights of the past, in order to be aware of the dangers and the unforeseen side effects of science,

---

[5] Appeared in *Horseless Age*, "a popular magazine for automobile enthusiasts" published between 1895 and 1918; from Ann Norton Greene, "Horses at Work: Harnessing Power in Industrial America," p. 261.

technology, and progress. While accepting, or even welcoming the inevitable movement forward, it would be wise if it is accompanied by soulful reflection. The present frenzy in experimenting with self-driving cars, which in a decade or two may dominate traffic in the United States, emphasizes the greater safety compared to the human driver. Death and injury due to car accidents will thus be considerably reduced, say those who favor this development. But this does not account for the alternatives, such as better education of drivers, law enforcement, or improvement and innovation in public transportation, reducing the crowding of streets and highways.

Unlimited rationalism is an oxymoron; axiomatically, the rational is rationed, an allocated portion, requiring a proportionate relationship. When truly rational, we rely on good reason and reasonable limitations, aware of our limited share in the world. However, when *one-sidedly* rational, we are blindfolded. And when we lack the link to the past that memory ensures, we fail to see the shadow-side of existence, whereby we invite the danger of cold evil, the evil of the fantasy of *perfection*, which always requires a condemned other, who carries the projection of the split-off shadow, as we have seen in the horrors of modern history.

In unequivocal faith in technology and uncritical confidence in the facts of science, the human soul and soulful reflection are lost. Jung writes, "... the soul is the magic breath of life. ... Being that has soul is living being. Soul is the living thing in man, that which lives of itself and causes life."[6]

The tradition of belief in three souls, body, animal, and spirit soul, each contributing to the growth and development of the whole, points to the need for human attention to participate in all three of these realms. Soul is psyche. And not only do our bodies have a soul—our breath of life—but we are also contained within Earth's soul, *Anima Mundi*, our planet as a living, breathing organism that is not separate from us but remains in a relationship of reciprocal care. Jung's prophetic statement that man's psyche must

---

[6] CW 9i., par. 55, 56.

be studied because it is the cause of all future evil rings ominously like a mountain bell across the valley.

In this book, I try to outline some of the human and psychological problems that we encounter in this new era of exponential scientific and technological progress. The idea of the Golem, the man-created humanoid machine that turns against its creator, stands at the center of our no-longer far-fetched imagery; by a gigantic leap the Golem quickly steps out of legend and fantasy into postmodern reality. Just like the Golem, new ideas and innovations are conjured up in the imagery of the human mind, now with augmented likelihood of actual implementation. Consequently, I will explore the changing character of the relationship between us humans and the image, and the dramatic impact this has in postmodern culture.

Image and image-formation are at the center of the human soul. "... [E]verything of which we are conscious is an image, and that image *is* psyche," says Jung.[7] Image-formation is a crucial aspect of the process of psychization, which is a primary process in human development and is essential to human self-healing, the significance of which is evident in the valuable phenomenon of placebo.

Thus, Part I of this book outlines a psycho-mythological development of the image, of the relationship between human and image, and the internalization of the image.

Part II is a reflection on the horrors of the perfection of the machine and the transition of the machine into becoming the master.

Part III focuses on the significant changes that take place when the image becomes externalized, when we no longer rely on our own internal, subjective memory and psychological capacities, but hand over cognition, memory, and image-formation to the machine. This part also entails a description of the elements of illness of our time, particularly the condition of transiency and the concomitant Transient Personality.

In Part IV, finally, ending on a hopeful note, the compensatory remedies of restorying the Self will be outlined.

---

[7] C.G. Jung, 'Commentary on "The secret of the golden flower", CW 13, par. 75.

PART I
Extracting the Image

CHAPTER

The Fish and the Fisherman:
The Birth and Development of the Image

Ever since I was a child in the 1950s, I was gracelessly uninterested in those endless, suffocating museum exhibits of ancient, Bronze Age and Iron Age tools. The dusty, murky, and smelly shelves of broken pottery, and long, boring depictions on bleak and yellowed cardboard, did not evoke my curiosity. I failed to become excited over the, to me, indistinguishable differences between rectilinear and curvilinear patterns of tools from different eras. Rather than making the ancestors come alive, I felt the past brought death into the present.

Thus, years later, Lewis Mumford's 1969 book, *The Myth of the Machine,* appeared as a welcome relief. He writes:

> Man's brain was from the beginning far more important than his hands ... ritual and language and social organization which left no material traces whatever, although constantly present in every culture, were probably man's most important artifacts from the earliest stages on; and that so far from conquering nature or reshaping his environment primitive man's first concern was to utilize his overdeveloped, intensely

> active nervous system, and to give form to a human self, set apart from his original animal self by the fabrication of symbols–the only tools that could be constructed out of the resources provided by his own body: dreams, images and sounds.[8]

Now that made sense! Dreams and images, rather than tools and stools. Jung says, "Whoever speaks in primordial images speaks with a thousand voices," and "In each of these images there is a little piece of human psychology and human fate, a remnant of the joys and sorrows that have been repeated countless times in our ancestral history ..."[9] Henry Corbin adds, breathtakingly and brilliantly, that while images take shape within our individual psyches, the image is not only within us, but we are also within the image.[10] "I dream and experience my dreams as inside me and yet at the same time I walk around in my dreams and am inside them," says James Hillman.[11]

We reside within an image of the world, and we relate to the world according to the images, ideas and perspectives that we have developed, and to a large extent, according to the views and the spirit of our times. We might, for a long period in history, reside within an image of the world that tells us that the Earth is the center of the universe, religiously convinced that this is the case. Or, we might be medically confident that bloodletting will cure even an ill president.[12]

The images of interiority cannot exist without the idea of psyche and soul. In soullessness, in fundamentalism, and totalitarianism, there are no images. The existence of the soul, that elusive, purely

---

[8] Lewis Mumford, *The Myth of the Machine*, p. 14.

[9] C.G. Jung, On the Relation of Analytical Psychology to Poetry, CW 15: par. 129, 127.

[10] Tom Cheetham, 2003, *The world turned inside out: Henry Corbin and Islamic mysticism*. Woodstock, CT: Spring Journal Books, p. 71.

[11] James Hillman, *Re-Visioning Psychology*, New York, p. xvi. Harper Perennial, 1992, p. 23.

[12] George Washington died in December 1799. About 40 percent of his blood was taken out over a 12-hour period. The medical practice of bloodletting continued into the 20th century.

poetical idea of *anima*, whether in man or woman, cannot be bound by earthly empires—neither by imperial rules nor by imperatives. "The inner attitude, the inward face," says Jung, "I call the *anima*."[13] The soul can only be poetically imagined, for instance as the image of a mirror that reflects the many faces of the images that arise in our psyche. As James Hillman so poetically says:

> Though I cannot identify soul with anything else, I also can never grasp it by itself apart from other things, perhaps because it is like a reflection in a flowing mirror, or like the moon which mediates only borrowed light. But just this peculiar and paradoxical intervening variable gives one the sense of having or being a soul. However intangible and indefinable it is, soul carries highest importance in hierarchies of human values, frequently being identified with the principle of life and even of divinity.[14]

The *idea* of the image constitutes the imperceptible nuclear force around which the images of soul and psyche gravitate. Hillman places imagination as the essential kernel of soul, by which he means "a perspective rather than substance," and individuality: "[T]he soul's first freedom [is] the freedom to imagine. This is the source of our peculiar individualness ..."[15]

Idea, imagination, and soul are intertwined. The capacity of symbol-formation and generating images are the insignia of the soul and the Self.[16] By means of this faculty, the original

---

[13] CW 6, par. 803. This is, of course, an interesting statement, reflecting an additional dimension to the well-known definition of anima as the feminine in the male psyche. Furthermore, the inward face applies equally to the female psyche.

[14] James Hillman, *Re-Visioning Psychology*, p. xvi.

[15] Ibid., pp. xvi and 39.

[16] In Analytical Psychology, the Self, often capitalized though Jung himself did not do so, indicates a psychic center beyond conscious identity (of which the ego is the center). It is both the center and the totality of the psyche, the archetype of meaning and wholeness. See C.G. Jung, "The self," CW 9ii., pp. 23-35.

imagelessness, for instance of the God-image or the archetypes, takes shape and receives contours and flesh.

The soul reaches down into the depths of imagelessness in contrast to externally generated images. In their discussion of Neumann's manuscript *Jacob and Esau*, Jung writes that "the 'imagelessness' [here referring to the God-image] is exceedingly important for the free exercise of intuition that would be prejudiced by a fixed image, and thereby rendered unusable."[17] That is, there is a danger that the image stifles into literalness, when fixed as an external depiction—which does not mean that we can do without images taking shape in external reality. On the contrary, we need that as much as we need the images of interiority. They are complementary.

We may then say that the soul is the place and the prism of imagination. The soul that resides in the individual is a prism that collects and gathers images brought to us from the objective psyche, the conscious and the unconscious aspects beyond our separate individuality.

These images may appear to us in distinct shapes, as often happens in a well-remembered dream. But the deeper we go, the more abstract are the images, until we reach the threshold of imagelessness. We might say that the divine resides in the imageless kernel of the image. In the process of becoming conscious, the original, archetypal imagelessness takes shape and appears as increasingly experience-near images. The more mystical aspect of existence pertains to the search for the imageless reflection behind the visible image.

---

[17] Jung, C.G. & Neumann, E. (2015). *Analytical Psychology in Exile: The Correspondence of C.G. Jung and Erich Neumann*, edited and with an introduction by Martin Liebscher, p. 56.

Aion and Chronos

In the course of history, as well as across societies, the place of image and psyche in relation to body, mind, and world have undergone changes. We might say that today psyche is increasingly becoming "brain," and image is increasingly computerized.

With his book *Aion*, which belongs to his later works, Jung intended "to throw light on the change of the psychic situation within the 'Christian aeon.'"[18] This is the era that corresponds to the age of Pisces, the approximately 2000-year period that is now coming to an end, as we enter the Platonic month of Aquarius, the water-carrier.[19] Jung writes, "If as seems probable, the aeon of the fishes is ruled by the archetypal motif of the hostile brothers, then the approach of the next Platonic month, namely Aquarius, will constellate the problem of the union of opposites. It will then no longer be possible to write off evil as the mere privation of good; its real existence will have to be recognized."[20]

Aion means a very long period of time or even eternity. It was thought of as an inner water, as Edward Edinger tells us, and in some Bible translations, the term aion replaces "for all time" or "forever."[21] In Homer, *aion* is comparable to *soul*.[22]

Aion can be compared to *Chronos*, which refers to the primordial god of time, past, present, and future, who devours everything living in due time.[23] Paraphrasing old proverbs, we

---

[18] Jung, CW 9ii, p. ix.

[19] The Platonic month is based on the precession of the equinoxes (i.e., when day and night are approximately equal all over Earth) and is supposed to span 2160 years.

[20] C.G. Jung, CW 9ii, par. 142.

[21] Edward Edinger, *The Aion Lectures: Exploring the Self in C.G. Jung's Aion*, Toronto, Inner City Books, 1996, 16, 17.

[22] Onians writes, "... elsewhere in Homer aeon clearly is not a period of time but a 'thing' ... persisting through time, life itself or a vital substance necessary to living." R. B. Onians, *The Origins of European Thought*, p. 200.

[23] Cf. Jung, CW 9ii, ¶215. Furthermore, the spelling of the Titan god's name varies. In the *New Larousse Encyclopedia of Mythology*, Twickenham, England, Hamlyn, 1987, his name is spelled Cronus (*Larousse*, 83, 90-91). Robert Graves as well, in *The Greek Myths* (Harmondsworth, England, Penguin, 1960, 39ff., spells his name Cronus, as does Edward

might say that while "Man thinks *time* passes, Time knows that it is *man* that goes." In "The Paradox of Time," Henry Austin Dobson succinctly says, "Time goes, you say? Ah no! Alas, Time stays, we go."[24] Chronos is frequently confused with Cronos, the Titan god who married his mother/sister, Rhea Kronia, Mother Earth in her capacity as Mother Time. She brings forth the eternal flow of generations and then consumes the children she brings forth, just as Cronos does later to avoid being supplanted by the younger generation. We might then, somewhat simplified, say that *Cronos* represents the time of ego, measurable, linear time, while *Aion and Chronos* reflect the time of soul and Self, beyond the boundaries and the limitations of ego consciousness.

In *Aion*, Jung moves from consciousness to the greater unconscious, from the earth of ego through the night of shadow to the waters of the Self in a spiral or circumambulatory progression. He clarifies his fundamental opus when, in his conclusion, he says, "In the end we have to acknowledge that the self is a *complexio oppositorum*, the opposites united in one image, precisely because there can be no reality without polarity."[25] Likewise, life itself depends upon the tension between the opposites, between growth and decay. Paradoxically, the tension between the extreme opposites

---

Tripp (*Handbook of Classical Mythology*, New York, NY, Meridian, 1970, 177-178). Jung variously spells his name *Chronos* (CW 8, 9ii) and *Kronos* (CW 5, 10, 11, 14). Barbara Walker (*The Woman's Encyclopaedia of Myths and Secrets*, San Francisco, CA, Harper Collins, 1983, 187) mentions, that the god often was "confused with Chronos, 'Time,' because Time swallows up everything it brings forth—actually a characteristic of Cronus's mother-mate, Rhea Kronia, the Goddess personifying time and fate." [While Rhea is daughter of Gaea, Mother Earth, and thus *sister*-mate of Cronus, she can also be seen as the triple goddess of the earth, thus embodying Gaea, Rhea, and Hera.] A representation in Greek mythology of sun-related time is *Chronos*, one of the seven winged horses pulling the sun-god Helios' golden chariot across the sky (cf. Larousse, 139). Following Jung in *Aion*, however, thereby not distinguishing the spelling of the Titan god's name from Helios's horse, his name is here spelled *Chronos* and as such. is not a direct representation of Time. Instead he epitomizes the aspect of time that is destructive and devours man.

[24] Austin Dobson, The Paradox of Time, in *Old-world idylls and other verses*, p. 175f, London: Kegan Paul, Trench, Trubner & Co., 1883.

[25] C.G. Jung, CW 9ii, ¶423.

of Eros, the life principle, and Thanatos, the death instinct, is necessary for life. Life does not exist, does not even come into being, without its opposite of death. Walter Benjamin writes:

> Today people live in rooms that have never been touched by death—dry dwellers of eternity; and when their end approaches, they are stowed away in sanatoria or hospitals by their heirs. Yet, characteristically, it is not only a man's knowledge or wisdom, but above all his real life—and this is the stuff that stories are made of—which first assumes transmissible form at the moment of his death.[26]

In the era of Pisces, the two fishes reside in the great waters, while in Aquarius the water has been rounded in and gathered into the bucket, held by the one human, the water-carrier. "In our era the fish is the content; with the Water-pourer, he becomes the container. It's a very strange symbol."[27] So in a way, they are each other's mirror-images, one *containing* the other. When looked at only from the point of view of digital, linear time, the images of Pisces and Aquarius become sequential, one follows the other. However, as each other's mirror-images, they compose a *complexio oppositorum*, a complexity or union of opposites, wherein each image *reflects* the other.

The Fish in the Water

Life in water precedes life on land. Not only the Chinese say that all life comes from the waters; in the Babylonian epic *Enuma Elish*, only water existed before the formation of heaven and earth. The divine offspring was born from the union of Apsu and Tiamat, the masculine and feminine personifications of the primordial

---

[26] Walter Benjamin (2002), The Storyteller, in *Selected Writings, vol. 3*, p. 151. Edited by Howard Eiland and Michael W. Jennings. Translated by Edmund Jephcott, Howard Eiland and Others. Cambridge, MA, and London, England. Harvard University Press.
[27] C.G. Jung, *C.G. Jung Speaking: Interviews and Encounters*, p. 413. (Edited by William McGuire and R. F. C. Hull. (1977). Princeton, NJ: PUP.

waters.[28] In the beginning everything was like a sea without light, and thus water is the most maternal, *matritamah* in Sanskrit, says the ancient sacred Hindu knowledge of the Veda—Vedic culture itself, and its poetry—developing along the waters. The waters held that very first germ where all the gods come together, the one in which all worlds abide, says the Veda.[29] That is, the water is the primordial womb in which all creation incubates, from which even the gods develop and come alive. In dreams, and on the personal level, Freud says, "Birth is almost invariably represented by something which has a connection with water."[30] To Jung, drawing water represents "the symbolic act or experience of the archetype: a drawing up from the depths."[31]

Likewise, in Genesis, the primordial waters were dark and undifferentiated, "the earth was without form, and void; and darkness was upon the face of the deep,"[32] until God's wind, *ruah* (רוח)—that is, God's spirit and soul—moved and resounded over the waters. Thus, God divided the light from the darkness, but he also separated water from water. In Hebrew, water is *maiim* (מים), and Heaven is *shamaiim* (שמים), which means "water is there," but which can also be understood as "fire-water" (esh-maiim). God's voice that resounded over the waters separated the earthly waters from the heavenly, the *matter* of water from the *spirit* of water. In fact, the Hebrew word for God, *Elohim* (אלוהים), can also be conceived of as *El haYam* (אלהים), "God of the Sea." Initially the water is an undifferentiated *prima materia*, preceding the creation of consciousness, but with a creative, birth-giving potential, the liquid of the whole verification, as Plato says.[33]

Water is the source of life, irrevocably evident in desert countries such as Israel, where it never is to be found in abundance,

---

[28] Nahum Sarna, *Understanding Genesis*, p. 4.

[29] *The Rig Veda*, Penguin Classics, p. 25ff.

[30] Sigmund Freud, *Introductory Lectures on Psychoanalysis*, SE 15, p. 153. Jung adds, "To be born of water simply means to be born of the mother's womb; to be born of the Spirit means to be born of the fructifying breath of the wind." (CW 5, par. 334.)

[31] C.G. Jung, CW 5, par. 349.

[32] Gen. 1:2.

[33] Plato, *Timaeus and Critias*, Harmondsworth, England, Penguin, 1977. pp. 83-87.

often rather in severe shortage, causing aridness and infertility. Israel is a land where, as it says in Psalm 107:33, rivers turn into wilderness and springs of water into dry ground. A fruitful land that easily becomes barren. We must keep in mind, though, as the Psalm (107:34) say, that it is the "wickedness of its inhabitants" that may cause barrenness—often the wickedness of draining Mother Earth of her milk, depleting her of her honey, without enough consideration of her need for rest.[34]

Correspondingly, by consciousness and devoted affection, the land may flourish and deserts bloom. As I have mentioned elsewhere, in the English translation of Israel's Declaration of Independence, it says that pioneers "made deserts bloom" (*lehafriach schmamot*), while the Hebrew wording was altered by one Hebrew letter the night before independence was declared, May 14, 1948, so that the final text reads, rather, to "make souls [or spirits] blossom" (*lehafriach neschamot*).[35]

We find water as the birth-giving source of life and as healing in sacred rituals. The initiation and salvation through Baptism is prominent in Western culture. The purification through immersion in water is found, as well, in the Jewish custom of the ritual

---

[34] Exodus 3:8, "And I have come down to save them from the hand of the Egyptians, and to bring them out of that land to a good and large land, to a land flowing with milk and honey." The Hebrew word *zavat*, translated as 'flowing,' indicates the profound fertility of the Earth Mother (or Earth Goddess), referring to discharge of genital fluids.

[35] Erel Shalit, *The Hero and His Shadow: Psychopolitical Aspects of Myth and Reality in Israel*, Hanford CA, Fisher King Press, 2011, 53. The question has not been resolved, whether the change was due to a Freudian "typing slip" or to Ben-Gurion's intimate knowledge of the Bible, e.g. Ezekiel 36:34, where land/desert and soul/spirit interchange; the English translation renders the verse "and the desolate land, after lying waste in the sight of every passerby, shall again be tilled," while the Hebrew reads

וְהָאָרֶץ הַנְּשַׁמָּה, תֵּעָבֵד, תַּחַת אֲשֶׁר הָיְתָה שְׁמָמָה, לְעֵינֵי כָּל-עוֹבֵר׳.

The relevant paragraph of the Declaration of Independence reads as follows in Hebrew:

מתוך קשר היסטורי ומסורתי זה חתרו היהודים בכל דור לשוב ולהאחז במולדתם העתיקה; ובדורות האחרונים שבו לארצם בהמונים, וחלוצים, מעפילים ומגינים הפריחו נשמות, החיו שפתם העברית, בנו כפרים וערים, והקימו ישוב גדל והולך השליט על משקו ותרבותו, שוחר שלום ומגן על עצמו, מביא ברכת הקידמה לכל תושבי הארץ ונושא נפשו לעצמאות ממלכתית.

cleansing bath, the *mikveh* (מִקְוֶה). Just as John baptized Jews in the River Jordan, so the Jews returning from exile in Babylon were sprinkled in the river's water to cleanse them of their sins. Likewise, prior to incubation—sleeping and dreaming in a sacred place—the patient coming for treatment in the sanctuaries of Asclepius had a purifying bath.[36] It is the "immersion in the bath" that enables the bond of love that makes the soul come into being and which is the *vinculum*, the ligament or bond between spirit and body.[37] Legend tells us that the source of all waters lay either under the Foundation Stone beneath the Temple Mount[38] or that it was to be found under the Tree of Life in the Garden of Eden.

Water is the substance of the primordial sea of the unconscious, surrounding earth and ego, from which life springs forth. In what Erich Neumann calls "prehuman" symbols, "the Mother is the sea ... and the child a fish swimming in the enveloping waters."[39] In biblical mythology, the waters is the place where God created living creatures; God first "created the great crocodiles, and every kind of creature that live in the waters" (Gen. 1:21). Likewise, life originated and evolved from the depths of the water.

Water is the transitional medium between the realms of the divine and the human; and in psychological terms, from the unconscious into consciousness. As Neumann adds, "The primal ocean ... is the source not only of creation but of wisdom too."[40]

Water, like feeling and emotion, is an agent of change and movement, as it sets us in motion. Water causes matter to dissolve, and like feeling, it can make hearts of stone melt. Where there is no water, no moisture, we cannot even cry. Poseidon, god of the sea

---

[36] C. A. Mayer. *Healing Dream and Ritual: Ancient Incubation and Modern Psychotherapy*, Einsiedeln, Switzerland, 1989, p. 69.

[37] C.G. Jung, CW 16, ¶454.

[38] Zev Vilnay, *Legends of Jerusalem*, Philadelphia, Jewish Publication Society of America, 1973, 8.

[39] Erich Neumann, *The Origins And History Of Consciousness*, Princeton, NJ, Princeton University Press, 1970, 43.

[40] Neumann, 23.

and ruler over fish (and son of Cronos), is called the earth shaker—everything may be shaken when we are flooded by emotions.

The fish is the living contents of the Great Mother matrix of the sea, of the unconscious part of the psyche, with which we may be out of contact. In order to retrieve what has been lost, and to find rejuvenation in the unconscious, the hero may have to dwell like Jonah in the belly of the fish. In some fairy tales, such as "The Fish and the Ring," the fish will serve as a mediator, bringing back the golden ring.

In astrology, Pisces is the 12th sign of the Zodiac, symbolized by two fishes whose tails are bound together by a wavy band, one swimming toward the spirit, the other toward matter, marking both an end and a new beginning. It is, quite naturally, a water sign. The fish in the sign of Pisces is prominently associated with Christ and Christianity. Jung considered the fish symbolism that gathered around the figure of Christ as synchronistic with the dawning age of the fishes.[41] *Ichthys*, the Greek word for fish, also serves as an anagram for "Jesus Christ, Son of God, Savior," and as such, Jung holds, "[T]his name referred to what has come up out of the depths. The fish symbol is thus the bridge between the historical Christ and the psychic nature of man, where the archetype of the Redeemer dwells."[42]

That is, Jung points at the fish serving as a bridging symbol between historical data and psychic image and representation. Likewise, the biblical story of creation says, "And God blessed them," referring to the fish, humans, and the seventh day, the Sabbath (Genesis 1:22, 1:28, and 2:3). Thus, fish, human, and the Sabbath are connected and related to one another.

The Sabbath is not only a day for human rest, but, as the biblical creation story tells us, it is a day that God rests. Thus, the Sabbath is no less a day to marvel at the wonder of existence and to *let the world rest*, than it is to interfere in the workings of the

---

[41] Edinger, 65.
[42] Jung, CW 9ii, ¶285.

universe by creating something new. The Sabbath, in distinction from the first six days of the week, reflects God's time, i.e., the *aion*, rather than human working hours, and signifies "the everlasting covenant" with the divine.[43] The Sabbath is supposed to be a foretaste of the Messianic Age, when the righteous will feast upon the leviathan, the huge, monstrous fish.[44]

In Hebrew "fish" is *DaG* (דג), a two-letter word (since Hebrew is written with consonants only). In Hebrew each letter has a numerical value, which for *DaG* (דג) is four and three, similar to the combined movement of three (biblical fathers) and four (biblical mothers), constituting the cycle of seven.

"Seven," *shev'a* (שבע), is the basis for the word *shavu'a* (שבוע), week. It is also found in the *shiv'ah* (שבעה), the seven days of mourning, and in *save'ah* (שבע), to be satiated, satisfied, as God was on the seventh day. It is also the root of taking an oath, *shvu'ah* (שבועה). Seven constitutes the cycle of divinity and humanity brought together, and the Sabbath constitutes the *hieros gamos*, the sacred marriage between the masculine and the feminine within the divine, between the god and the goddess. On the Sabbath, claims the mystic, he is penetrated by the divine phallus and becomes the female counterpart in his relation to God.[45]

So, we have reason to conclude that the image of the fish embodies a link between God and man. The fish in the water carries the elements of the Self, of divinity. "Alchemists imagined the fish eyes as shining sparks within primal matter that intimate ever-aware multiple luminosities within the dark waters of psyche."[46] Like the patriarchs, they are closer to God, and serve,

---

[43] Exodus 31:16: "The Israelite people shall keep the Sabbath, observing the Sabbath throughout the ages as a covenant for all time."

[44] Isaiah 27:1: "In that day the Lord will punish, with his great, cruel, mighty sword Leviathan the Elusive Serpent—Leviathan the Twisting Serpent; He will slay the Dragon of the sea." That is, the monstrous leviathan, embodying chaos and evil, will be destroyed at the end of time (The Jewish Study Bible, p. 817).

[45] Moshe Idel, Sexual metaphors and praxis in the Kabbalah, in Mortimer Ostow, *Ultimate Intimacy: The Psychodynamics of Jewish Mysticism*, London, England, Karnac, 1995.

[46] Ami Ronnberg, (ed.), *The Book of Symbols*, p. 202.

like the Sabbath, as intermediary between the divine and the human, an archetypal potential that has to be extracted from the waters in order to be realized. And like the eyes of God, the fish's eyes never close.

Moreover, the fish serves as vessel for the hero's transformation, as for instance in Jonah and the whale-fish and other similar night-sea hero journeys. This further endorses the designation of the fish as a link between human and the divine, bringing elements of the divine into the realm of the human—which is the hero's task as well.

In Pisces, the containing vessel, the fish, is itself contained in its medium, the water. Considering the implications of our time, as will be reviewed later, we must be astutely aware that in Aquarius, however, it is the other way around—the water is contained in the vessel.

The Fisherman

The fish and the fisherman are intimately connected with each other. They may, in fact, be two separate aspects of the same thing, the extractor and the extracted. Jung points out how "anyone born under Pisces may expect to become a fisherman or a sailor, and in that capacity to catch fishes or hold dominion over the sea."[47] He mentions the Babylonian culture-hero Oannes, who himself was a fish, and particularly Christ as fisher of men; and we may add the Phoenician god Dagon, which means "strength of the fish." Half-fish and half-man, he was a teacher of humankind, arising from the sea each day, returning at night. Christ is the fish that is Eucharistically eaten, so that by incorporation he may be remembered—as he says "in remembrance of me"[48]—and thus one who eats the Eucharist is united with the spirit of divinity conveyed by Jesus. In remembrance, in creating wholeness by "re-membering," by "gathering the members," lies redemption. Christ

---

[47] C.G. Jung, CW 9ii, ¶174.
[48] Luke 22:19; 1 Corinthians 11:24.

is also, says Jung, "the hook or bait on God's fishing-rod with which the Leviathan, the primordial sea-monster—death or the devil—is caught."[49] In the Scriptures, Moses (משה), the Lawgiver, spent 40 days on the mountain to receive the stone tablets with the Ten Commandments from God to bring to the people. Hence, like the fish, a mediator and intermediary between God and human, Moses was himself drawn out of the water, which is the very meaning of his name.

It is the fisherman who fishes the fish, who pulls it out of the waters and brings it for man to eat, that is, for the human to incorporate its soul-substance. "Fishing is an intuitive attempt to 'catch' unconscious contents (fishes)," says Jung.[50] The fish that is drawn out of the sea is an element of the divine that can be incorporated, internalized, eaten by humans. But the divine soul and spirit are conveyed to man in the disguise of, as Jung says, the fish's "opprobrious qualities."[51] That is, we humans cannot rid ourselves of the shadow. I will here leave aside the question of the shadow of the God-image, though the Self would not be able to execute its task as the central archetype of order and meaning, were it not bipolar, entailing its own shadow, thus being, a *complexio oppositorum*. In fact, when the Self is elevated to an idealized shadowless God-image, as if all good, it all too easily compounds as a reality-projection, for instance as the worshipped king or leader, whereby the Self may turn into its shadow of cultism or fascism. Archetypal identification with an idealized, one-sided, supposedly shadowless Self-principle is the basis of fundamentalism, as shall be expanded upon later.

The fisherman reaches down into the depths of the unconscious. He may be taken as an image of the soul's very extraction from the divine, of human psychic development, of separating the soul and the image from the gods, and bringing it

---

[49] C.G. Jung, CW 9ii, ¶174.
[50] C.G. Jung, CW 9ii, ¶237.
[51] Jung, CW 9ii, ¶174.

into the individual human psyche.[52] The elements of the Self that may be extracted as fish constitute the *ruach*, the spiritual soul, whether we are fishing in our interior waters, or in the *anima mundi*, the world soul.[53]

Ensoulment, Image, and Dreaming

Fishing is a silent activity; it is all too easy to scare the fish away. The fisherman sets sail when still dark, at dawn, before the light and the sounds and the noise of the day break the silence of the night, when the soul can still be heard. This is the time of early morning dreaming, when we wake up, sometimes to realize that we are dreaming. The fisherman silently throws his net in the hope of catching at least a glimpse of the Fish's Eye, always wide open, the way we might say that the Self, as the God-image in our psyche, always has its eyes open, ready to gaze, the gaze of the Self, at the fisherman, like a mirror of the soul. The fisherman pertains to the threshold of psychic life and the very first stages of germinating consciousness. The image of the fisherman reflects the psychological hero who ventures into the darkness of the mind's night, bringing pearls from its treasure house into ego consciousness.

Marie-Louise von Franz says, "It is sometimes revealed very clearly to us that [the creation myths] represent unconscious and preconscious processes which describe not the origin of our cosmos, but *the origin of man's conscious awareness of the world.*" [Italics in original.][54] In Egyptian creation mythology the Sun God, consciousness, rises from Nun, the primordial ocean, in which the

---

[52] In Plato's symposium, Diotima explains to Socrates that the spirit (daemon) is midway between the divine and the human, acting "as an interpreter and means of communication between gods and men" (Plato, *Symposium and the death of Socrates*, 202 d-e, p. 36). Heraclitus, however, promotes the daemon into the human psyche, "A man's character is his daemon" (Heraclitus, *The Fragments of Heraclitus*, translated by G. T. W. Patrick, p. 112, fragment 121.

[53] Cf. Gershom Scholem, *On the Kabbalah and its Symbolism*, pp. 193, 194. The Chinese differentiated between *p'o*, the material soul, and *hun*, a spiritual soul.

[54] Marie-Louise Von Franz, *Creation Myths*, Boston, MA, Shambhala, 1995, 5.

germs of every thing and every being dwell before creation.[55] The light of human consciousness, which Neumann calls a "small and late child of the unconscious," rises, emerging from the waters.[56] Like the unconscious, the waters hold the phenomena of the world in germinating states of undifferentiation.

When the world is created, as it is said, by "the *word* of God's mouth," it is quite evident that we are dealing with the creation of conscious awareness rather than actual, physical existence. Concomitantly, it indirectly pertains to the creation of the God-*image*, the image of a God that creates the world by means of the *word*, or by means of a story, a story of creation.

There is then no longer a preexisting *matter*, but a nothingness, a no-thing-ness, that pertains to soul, mind, psyche, spirit, imagination. "The God of Jewish myth," says Angelo Rappoport, "is clad in clouds and visits His worlds upon the wings of the wind."[57] *Wind* in Hebrew is *Ruah*, which also means soul/spirit. According to a Jewish myth, the Creator produced, from complete nothingness, "a fine subtle matter, which had no consistence whatsoever, but possessed the potential power to receive the imprint of form."[58] Could there be a more lucid definition of what we imagine the "archetype-as-such" to be?[59]

The creation myths bring man from the nothingness and chaos of eternity into the confines of experienced time, Chronos, which establishes the ego and separates the individual from the world around him or her. At the dawn of history, the human

---

[55] *Larousse*, p. 11.

[56] Erich Neumann, *Jacob and Esau: On the Collective Symbolism of the Brother Motif*, p. 77.

[57] Angelo Rappoport, *Ancient Israel: Myths and Legends*, London, Senate, 1995, 3.

[58] Rappoport, 5.

[59] The "archetype-as-such" has no consistency ("no material existence of its own," says Jung). It is empty, but holds a "possibility of representation." As I see it, the archetype (as such) receives form only when related to ideationally, and then becomes what Jung calls an archetypal image (such as "the wise old man"). The archetypal image, then, imprints on the personal psyche in such a way that complexes are constellated (in conjunction with the internalization of external objects). The complexes give visible and com-prehensible shape to underlying archetypal blueprints.

emerges from a state of oneness with the rest of nature.[60] In this original state, there is no separate identity, the ego has not yet differentiated from the Self. There are no ego-boundaries, and man is at this time hardly aware of his own separate existence. He has, says Jung, "a minimum of self-awareness combined with a maximum of attachment to the object; hence the object can exercise a direct magical compulsion upon him."[61]

In this early state, a feeling of identity between the individual and the collective predominates, and thus, the human soul was not necessarily located inside the body but could as well be found in nature, outside the body-boundaries. This is what Jung, following the anthropologist Lèvy-Brühl, calls *participation mystique*,[62] and Melanie Klein calls *projective identification*.[63]

The first traces of differentiation might have been pain and touch, then auditory—a crack in nature's wholeness by a squeak here, a chirp there, a twitter somewhere, a rattle elsewhere. Then the frightening howling of the wolves—and as sometimes happens in the process of psychization, man's pathology echoes introjected nature—just as "the gods have become diseases," as Jung says[64]— and so the howling returns in the coprolalia of Gille de la Tourette; and then the lion's roar; a rough and agitated wind, and the thunder of Thor's chariot; and, in Ezekiel, God's voice was like "the sound of many waters."[65] Then, ascending to the sky from within

---

[60] Erich Fromm, *The Sane Society*, New York, Fawcett Premier, 1965, 29ff.

[61] C.G. Jung, CW 8, ¶516.

[62] C.G. Jung, CW 6, ¶781.

[63] Cf. Jean Laplanche and J.-B. Pontalis, *The Language of Psychoanalysis*, London, England, Karnac Books, 1988, 356, and Betty Joseph, Projective Identification: Clinical Aspects, in Joseph Sandler (ed.) *Projection, Identification, Projective Identification*. Karnac, 1988 pp. 65-76.

[64] "The gods have become diseases; Zeus no longer rules Olympus but rather the solar plexus, and produces curious specimens for the doctor's consulting room," in 'Commentary on "The Secret of the Golden Flower", CW 13, par. 54.

[65] Ezekiel 43:2. Tourette syndrome reflects an interesting development: it was once considered rare, and included coprolalia, the compulsion to verbally utter obscenities. The scope of the diagnosis has broadened and now encompasses 1 percent of school-age children and adolescents, without the requirement of coprolalia. Does this mean that

the cave, emerging from the mother's womb, we hear the cry of a newborn baby, the birth of the human.

Following the grand slumber in the oneness of the womb of the Great Mother, the human might have come to dwell in a dream-like state that we return to every night, no longer unconscious, but also not having awoken yet to full consciousness. *This is the dream state, which shapes the beginning of man's conscious and cultural development.* The human soul separates from the body of nature and from the nature of the body. This is the beginning of psychization, as Jung calls it,[66] or ensoulment, the separation of the archetypes from the world of the instincts, the birth of soul.

Psychization is the process whereby aspects of instinctual phenomena are transformed into conscious experience. Dreaming —not necessarily the recorded and remembered dream itself—is instrumental in the process of psychization. Let me clarify: In ontogenesis, the development of the individual, we know that the baby sleeps approximately 16 hours a day, dreaming around 50 percent of that time, or in the premature baby, up to 75 percent. That is, dreaming is the infant's main activity. We may wonder, "But what does the newborn child dream about? It has not yet experienced anything!" A question hard to answer, when we think solely in terms of development and experience. It is easier from an archetypal perspective, whereby the *idea* might precede the actual *experience*, archetype predates ego, *imagination* comes before the actual *image*, and the process of dreaming precedes the consciously remembered dream. The baby's *dreaming* transforms a touch, a pressure, a pain and a pleasure, a sound and a sight into the capacity to *experience* the touch, the pain and the pleasure, the sound and the sight.

---

the syndrome is now more correctly diagnosed, or that the diagnostic criteria have been expanded, so to include what was previously not considered rendering a psychiatric diagnosis?

[66] Jung, CW 8, ¶234: "Instinct as an ectopsychic factor would play the role of a stimulus merely, while instinct as a psychic phenomenon would be an assimilation of this stimulus to a pre-existent psychic pattern. ... I should term it *psychization.*"

The beginning of REM sleep and dreaming, supposedly about 130 million years ago, marks a major transition in human evolution.[67] The *process* of dreaming, even before the story or the *content* of the dream, provides for a beginning differentiation between physical experience and its psychic *re*-presentation. Dreaming indicates and promotes human psychic development, phylogenetically as well as ontogenetically. By means of early dreaming, experience is presented again, that is, represented in the psyche, thus enabling the human infant to *relate* to experience. This would mean that the *image* of the experience is extracted from somatic experience, similar to the distillation of essence from matter in homeopathic medicine.

For example, a middle-aged man with a tendency to somatize his psychological and emotional condition, and neglect his body, dreamed that he beats up his sister.[68] He *has* no sister, nor did he ever have a close relationship with a woman, or a man for that matter, because his anima, the essence of relatedness, had not yet woken up. His sister-*anima* had, so to speak, taken unmediated possession of his body and merged with it because he was unrelated; relatedness requires differentiation, an *other*, even the differentiation between psyche and soma.

By raising the anima into the psyche, he could slowly relieve his body from carrying the entire burden of the unintegrated anima that had remained glued to his beaten body. The anima as a *psychic image* could take shape. While still under the spell of his pathology, he was initially able only to beat her up (in his dreams). He tried to destroy her, to put her back into his body, so that he could continue to neglect her and avoid relating to her, his anima, the essence of relatedness. Only much later, when he would dream of, for instance, the woman in the bath, slowly emerging from the waters, comparable to an awakening consciousness, could significant development take place.

---

[67] Anthony Stevens, *Private Myths: Dreams and Dreaming*, p. 91.
[68] Erel Shalit, *The Complex: Path of Transformation from Archetype to Ego*, Toronto, Canada, Inner City Books, 2002, 32-33.

By raising the soul from the soma, extracting the fish from the waters, by the formation of an image through dreaming, the body could begin to heal. At a different level, closer to consciousness, and more classically ego-based, this is the idea of *healing by interpretation* —that the hysteria that has taken its grip upon the body be dissolved by being understood by the conscious ego. When the experience is de-somatized, it can be imagined and reflected upon. The infant who remains at a purely sensory level is undeveloped, uncommunicative, and unrelated, as if not fully human. We become human by means of soulful imagination, whereby "events deepen into experiences," as Hillman says.[69] And if we develop from touch to sound to sight, then notice—almost all dreams are visual; auditory experience is present in fewer than half of all dreams; and touch, taste, smell, and pain in a relatively small percentage of dreams. That is, psyche reverses somatic, bodily nature.

Soul and Reflection

Robert Stein writes in *Incest and Human Love* that "what made the barbarians barbaric in the eyes of the Greeks .. was the absence ... [of a] sense of proportion that grows out of *deep and constant reflection*." [My italics.][70]

The uncivilized and brutal barbarian does not know to reflect. Jung says:

> *Reflexio* means "bending back" and, used psychologically, would denote the fact that the reflex which carries the stimulus over into its instinctive discharge is interfered with by psychization. ... *Reflexio* is a turning inwards, with the result that, instead of an instinctive action, there ensues ... reflection or deliberation. ... Through the reflective instinct, the stimulus is more or less wholly transformed into a psychic content. ... Thus, in place of

---

[69] James Hillman, *Re-Visioning Psychology*, New York, Harper Perennial, 1992, xvi.
[70] Robert Stein, *Incest and Human Love: The Betrayal of the Soul in Psychotherapy*, Baltimore, MD, Penguin, 1974, 78.

the compulsive act there appears a certain degree of freedom ...[71]

The implication is that when we are unrelated and un-reflective, we become imprisoned by compulsive acting. There is no freedom and no reflection in compulsion, only machinelike repetition. Soul is the reflection in the mirror of the water.

When the ego is narcissistically inflated, we also do not reflect. If we do reflect, we have to renounce our narcissism. In narcissism, the ego is inflated—there is no separate *Other*: I see no other person; neither do I see gods in the elements, nor monsters in the shadow. In narcissism, we must not know who we are; if we do, we become dethroned from the emptiness and nothingness of our as-if supremacy. The river-nymph Liriope asks Tiresias the seer, if her son would live "long years and old age enjoy," to which Tiresias answers affirmatively, on condition that Narcissus, her son-to-be-born, "shall himself not know."[72] When he does come to know himself, falling in love with his self-reflection, he dies. The enchantment of youthful beauty shatters when Narcissus realizes that it is himself he sees, as he merges with his mirror-image, reflected in the water. Thus, when he begins to know himself, narcissism splits, transformed on the one hand into the shadow of death in the water, and on the other hand into a flower in the field.

In narcissism, we strive to induce the ego, I, *me*, with the spirit of godliness, sanctifying the ego rather than sacrificing its superiority. To a certain extent, in primary narcissism, this brings healthy self-esteem. In pathological narcissism we *do* strive for the great *spirit* and the burning *flame* and the divine *voice*, but the spirit becomes *spirit* or *alcohol abuse*, the tiresome flame becomes *burnout*, and the inner voice becomes an outer *Echo* chasing Narcissus, an empty repetition of our own words of self-aggrandizement. We neither hear nor see any other. In narcissism,

---

[71] Jung, CW 8, ¶241.
[72] Ovid, *Metamorphoses*. Translated by A. D. Melville. Oxford, Oxford University Press, 1986, p. 61.

the other, whether the inner Self or the other person, is exploited to reflect and to mirror a self-inflated ego. This contradicts the process of individuation, in which the Self "descends on the ego" and "imposes assignments ... the ego would much rather not do."[73]

To see an *other*, Narcissus must die. Only then can we bear to look into the mirror of the soul, which amounts to *reflection*. "True, whoever looks into the mirror of the water will see first of all his own face," says Jung, and continues:

> Whoever goes to himself, risks a confrontation with himself. The mirror does not flatter, it faithfully shows whatever looks into it; namely the face we never show to the world because we cover it with the *persona*, the mask of an actor. But the mirror lies behind the mask and shows the true face.[74]

That is, the mirror of interiority reflects not the outer mask or façade, but the true inner face.

Thus, narcissism dies at the very moment of *reflection*, which is the hydrargyrum, the "water-silver", the mercury of the soul. That is, reflection is the transformative element that enables human depth and consciousness rather than the inflation that comes with remaining bound to the godlike world of archetypes. While narcissism is a crucial element of the soul, it prevents soulfulness if it is not defeated at the moment of reflection. The soul is a reflective mirror, or, as the Sioux Indians say, "The moon is the Goddess's mirror reflecting everything in the world."[75]

The capacity for image-formation and reflection is essential to human psychic life, composing the soulful link of consciousness to the natural tendency of the unconscious to form symbols. Constituting the transition of the god-image into man's psychic life, dreaming, image-formation and reflection enable human inventiveness and further progress.

---

[73] Edinger, 87.
[74] C.G. Jung, CW 9i, ¶43.
[75] Walker, 669.

CHAPTER

From Craftsman to Computer

Just as the fisherman extracts soul as a reflective mirror and as an intermediary between divinity and the human, so the craftsman, such as the carpenter, brings archetypal patterns and blueprints into work in the human sphere. However, we must keep the words of warning in mind, "Cursed be the man who makes any engraved or molten image, an abomination to the Lord, the work of the hands of the craftsman, and sets it up in secret."[76]

These may be the words of a jealous Old Testament God, holding on to unmitigated power. Or they may be the words of a God who does not fully trust man, who all too well knows the creature he has brought forth. I think this warning refers to the combination of *engraved image*, *abomination*, that is, detestation or corruption, and *secrecy*. That is, humans might misunderstand the freedom given to them, and that they grab for themselves. Man might then, by overemphasizing an ego perspective, literalize the image, as Hillman says.[77] He may come to abominate and curse the grand *Other* and secretly conspire against the divine or the

---

[76] Deuteronomy 27:15.
[77] James Hillman, *Re-Visioning Psychology*, p. 48.

transcendent, trying to replace it by "Man, the Great Master," who in self-seduction becomes inflated, believing in his Godlike powers, losing his moral and ethical concerns. This happens when man recklessly and impudently extracts the soul and the spirit, whereby only unyielding, stiff matter remains. In contrast to the craftsman, who with patience and care draws from transcendent inspiration, the ego's attempt to take possession of the Self's resources breaks the connection with the imageless aspect of the image, with only its material concretization remaining.

Human imagination, by means of which we can *observe* and *relate to* experience, not only be the *subject* of experience, enables us to replicate the patterns and functions of creation. Besides the craftsman, the capacity to recreate has often been represented as well by the potter.

The Egyptian god Khnum, *The Molder*, was the potter, who on his wheel shaped the world-egg. With crafting hands, the potter shapes the clay into a pot, into an urn. Just as the body holds the human soul, the amphora—the tall, two-handled vessel that once was the bucket of the water-carrier, Aquarius—holds the water.

Neil Postman, the eminent American social philosopher, describes in his imperative book *Technopoly* how tool-use was followed by a cultural stage of development in which technological progress came to influence the culture of a society and an era. The craftsman, who enables the Self's unfolding in the human sphere, belongs to this early tool-using stage in human development. The technocratic stage of cultural development is now being replaced by a *technopoly*, defined by Postman as "totalitarian technocracy."[78] Man's inventions and technological progress have influenced the way we look at the world and how we position ourselves in relation to what we see. Postman expands the old saying that "to a man with a hammer, everything looks like a nail" and says, "To a man with a pencil, everything looks like a list. To a man with a

---

[78] Neil Postman, *Technopoly: The Surrender of Culture to Technology*, New York, NY, Vintage, 1993, 48.

camera, everything looks like an image. To a man with a computer, everything looks like data."[79]

A transition from prayer to God to material welfare took place when the mechanical clock was introduced. It had its origin in the Benedictine monasteries of the 12th and 13th centuries. The mechanical clock was to provide regularity of prayer and precision in man's devotion to God. But by the middle of the 14th century, it had "moved outside the walls of the monastery, and brought a new and precise regularity to the life of the workman and the merchant,"[80] serving the accumulation of money rather than devotion to God. "The mechanical clock," writes Lewis Mumford, "made possible the idea of regular production, regular working hours and a standardized product."[81] In the course of time, time itself became standardized. In 1884, due to the needs of the quickly developing railroads, time was organized into 24 global time zones. Thus "God's own time," as it was phrased, and solar time were replaced by ego-promoting order and differentiation. Just imagine railway schedules when time differed between adjacent villages and towns. Eventually, the skilled worker was replaced by the worker "who merely kept the machine operating," and by 1850, "the machine-tool industry was developed—machines to make machines."[82]

The pounding of progress caused industrious excitement, and the factory wheels turned ever faster. The masses had no alternative but to follow suit, whether willingly or reluctantly. There were, of course, those who saw danger: In *Jerusalem*, William Blake wrote of the shadows cast by the "dark Satanic mills" of the factories of the Industrial Revolution, forcing men to be enslaved in hard and soulless work,[83] and Emile Zola vividly portrayed the social ills of

---

[79] Postman, 14.

[80] Postman, 14.

[81] Lewis Mumford, *Technics and Civilization,* New York, NY, Harcourt Brace Jovanovich, 1963, 15.

[82] Postman, 42.

[83] William Blake, *The Complete Poetry and Prose of William Blake* (ed. David Erdman), Bantam Doubleday Dell, New York, NY, 1997.

prostitution, alcoholism, and spiritual emptiness[84] that come with the poverty of greed and gluttony.

Money was initially a great invention that easily may turn into self-generating growth, based on greed, speculation, and artificial pyramids when it is worshipped as the god of Mammon without awareness. The root of this word. Mammon, comes from *mamon,* which means, that which one trusts. Trust is at the core of money as a means for complex exchange of goods, values, and services, but how easily its shadow side takes control.

Bezalel and the Golem

Like everything else, development and progress are accompanied by shadow aspects. The fantasy of a shadowless future tends to create a dangerously "brave new world." While the inevitable shadows cannot and should not prevent development, they need to be considered and accounted for. This, in fact, is the main thesis of this book, a message that seems particularly important in times of unimpeded belief in the benefits of a technological future determined by bright fantasies that break up the basic integrity of the human being, creating cyborgs as swiftly as fishes that fly. We have reached a stage in which technology enables the implementation of our most wonderful as well as the most outrageous ideas that in the past were confined to fantasy and fiction. It seems, however, as if ethics and morals, psychology and sociology, all lag far behind the rapid transition from machines serving man to the human becoming a (vulnerable) link in the interconnected network of things and minds.

Turning the human into a servant of the machine, so eloquently illuminated in Chaplin's *Modern Times*, tends to eliminate the crafts-man. The successive developments of the machine in postmodernity, with the machine increasingly acquiring traits previously ascribed

---

[84] Cf. Emile Zola, *Nana* (trans. Douglas Parmee), Oxford, England, Oxford University Press, 1998, and Emile Zola, *L'Assommoir* (trans. Leonard Tancock), Harmondsworth, England, Penguin, 1972.

to human thought and behavior, obviously pose ethical questions and cause psychological dilemmas.

The craftsman and the man-created anthropoid, such as the *Golem*, stand opposing each other on the historic field of human character. The craftsman humbly pours soul and spirit into matter, transforming archetypal patterns of nature into the realm of human craft and creation. The Golem, on the other hand, is a result of an inflated ego that takes possession and attempts to replace divine nature, making man the Grand Creator.

In Greek mythology, Hephaestus is the master craftsman. In the Bible, the first to replicate creation was Bezalel, the craftsman who built the Tabernacle. In Exodus, God speaks to Moses and tells him, "I have called Bezalel and I have filled him with the spirit of God, in wisdom, and in understanding, and in knowledge, and in all manner of workmanship, to devise skillful works, to work in gold, silver, and brass, and in cutting of stones for setting, and in carving of wood, to work in all manner of workmanship."[85]

The Tabernacle was the tent set up by Moses in which the Ark of the Covenant—a chest of acacia wood holding the stone tablets of the Ten Commandments—was carried through the wilderness. Bezalel, whose name means "in the shadow [or 'in the image'] of God," "knew the combinations of letters with which heaven and earth were made" and thus was able to build the tent, which was considered "a complete microcosm, a miraculous copy of everything that is in heaven and on earth,"[86] i.e., an *imago mundi*, a crafted replica of the universe, wherein the divine can dwell on earth.[87]

What we might call "the craftsman ego" is rooted in the soul, enabling the inner Self to unfold in the realm of reality. It is worth noticing that the personal father of a hero, besides his divine father, is often a craftsman. "For the hero to set out on his journey into the depths and the vast lands of the unconscious, he or she needs to

---

[85] Exodus 31:1-5.
[86] Scholem, 167.
[87] See *The Cycle of Life: Themes and Tales of the Journey*, p. 120.

be ignited by the very sparks of the beyond as it manifests in ego-consciousness."[88]

The craftsman serves as a metaphor for the human ego's capacity to extract the images that arise from the inner depths, whether we call it inner waters, the soul or the Self, or the divine spark in the human. It is the craftsman who molds them into actual form in the world of reality. In contrast to the insincerity of imitation, the craftsman replicates the divine on earth by means of his skill, patience, carefulness, and hard work. In fact, in *The Guide for the Perplexed*, Maimonides,[89] the Rambam, refers to the craftsmanship of the Tabernacle as wisdom. We might say that craftsmanship brings wisdom alive; it is by means of the precise and exact craftsmanship that wisdom, Chochmah, and the Shekhinah can become manifest, thus anchoring the divine in earthly reality.

The man-made anthropoid, on the other hand, is the product of an inflated ego that has taken *possession* of the Self's resources. It is as if the human ego claims Godlike qualities and abilities without being aware of the limitations and restrictions that pertain to us as mortals. In contrast to the craftsman-ego, in which the Self unfolds, in which the patterns of nature serve as blueprints for what is being crafted, in the case of creating the anthropoid, there is no longer an ego in which the Self unfolds, but an overblown ego that claims total power and disclaims the soul and the significance of anything beyond itself. As a consequence, ethical concerns are sidestepped. In the postmodern world, the issue might be to what extent we can limit our human hubris and prevent the Golem, for instance in the shape of Artificial Intelligence, from rising against its creator. This is an ethical and psychological concern. If we discard the qualities of the craftsman to the dunghill of history, releasing ourselves from the bonds that bind us to existential limitations, we endanger the human species by creating the monsters that rise against us.

---

[88] Ibid, p. 117-122.
[89] Maimonides, *The Guide for the Perplexed*, III:54.

The idea of creating artificial, humanlike anthropoids is, as Gershom Scholem notes in his 1955 Eranos lecture on *The Idea of the Golem* "widespread in the magic of many people."[90] There is an ancient mystical Jewish idea about the creation of an anthropoid, or automaton, a man-made man, and in modern times we have, for example, Mary Shelley's novel from 1818 about Frankenstein's monster and the short story by Asimov, turned into the Robin Williams film *Bicentennial Man*.[91] We find the idea in Faust and Paracelcus's *homunculus*, little man, who was an "artificial embryo, for which urine, sperm, and blood [were] considered as vehicles of the soul-substance, [and] provided the prima materia."[92]

The best-known modern version of the golem is based on the legend of the 16th-century Rabbi Loew from Prague. This story was revived at the turn of the 19th century by several authors, notably Gustav Meyrink.[93] Jung refers to Meyrink's *Golem,* when he asserts that "the image of this demon forms one of the lowest and most ancient stages in the conception of God."[94] Meyrink uses the legend of the man-made creature as an allegory of how man is reduced to a robot by the pressures of modern life. The man-made golem is the hero's *double,* or *shadow*. It eventually turns out that it is the beggar—who in his outstretched hand, as a matter of fact, holds a distinct facet of our shadow—who is Meyrink's golem.[95] Incidentally, Meyrink himself had been a bank manager before turning to alchemy and Kabbalah. He was convinced that the *Philosopher's Stone* was to be found in the Prague sewer system— in spite of this all-too-literal understanding of alchemical symbolism, whoever has sensed the mystique of Prague must admit that if

---

[90] In: Gershom Scholem, *On the Kabbalah and its Symbolism*, New York, NY, Schocken, 1969.
[91] *Bicentennial Man*, 1999. Screenplay by Nicholas Kazan; Directed by Chris Columbus. Based on the short story by Isaac Asimov, in *The Bicentennial Man and Other Stories*, New York, NY, Doubleday, 1976.
[92] Scholem, 197.
[93] Gustav Meyrink, *The Golem*, Cambs, England, Dedalus, 1995.
[94] C.G. Jung, CW 7, ¶154.
[95] Meyrink, 184.

anywhere, the stone probably is to be found in the *nigredo*, in the darkness of Prague's shadow.

"The development of the idea of the golem in Judaism," writes Gershom Scholem, is connected with "the magical exegesis of the *Sefer Yetzirah* and with the ideas of the creative power of speech and of letters."[96] The *Sefer Yetzirah*, "Book of Creation" or "Book of Formation," is an ancient Jewish mystical treatise, a detailed cosmology, an account of the origin of the universe, "grounded in the assumption that combinations of letters are both the technique to create the world and the material for this creation."[97] In the *Book of Creation* we read:

> Twenty-two letters, He [God] engraved them and He carved them and weighed them and exchanged them and combined them, and He created by them the soul. ... Twenty-two basic letters [the Hebrew alphabet], fixed in the wheel, in the 231 gates [the number of pairs that can be computed from the twenty-two letters]. And the letters turn in a circle, in a wheel, forward and backward, and thus everything created and everything spoken emerge out of one name.[98]

The name referred to is the name of God—and God is often referred to as HaShem (השם), *The Name*—from which the universe evolves.

Among the many versions of the golem, the following text is based on an ancient inscription attributed to the Tannaite Rabbi Judah ben Bathyra, which means that it dates from the period after the destruction of the Temple in the year 70 C.E. to around the year 200:

---

[96] *Encyclopedia Judaica*, quoted in Moshe Idel, *Golem: Jewish Magical and Mystical Traditions: On the Artificial Anthropoid*, Albany, NY, State University of New York Press, 1990, xvii.

[97] Idel, 9.

[98] Cf. Ibid. p. 10, and Aryeh Kaplan (1997). *Sefer Yetzirah: The Book of Creation*, San Francisco: Weiser Books, pp. 108-136.

Ben Sira wished to study the *Book of Creation*. A heavenly voice [bat kol] came forth and said: You must not study alone. He went to his father, the prophet Jeremiah, and they studied the book for three years. They then set about combining the alphabets according to the secret principles of word formation, and a man was created to them, on whose forehead stood the letters YHWH (Elohim) Emeth (יהוה אמת), *God is Truth*. But this newly created man had a knife in his hand, with which he erased the aleph [first letter] from *emeth* [truth]; so that there remained *meth*, dead, that is, YHWH (Elohim) meth (יהוה מת), *God is Dead*.[99]

In another version, more similar to other ones, on the golem's forehead it just said *emeth* (truth), like on Adam's forehead:

Then the man they had made said to them: God alone created Adam, and when he wished to let Adam die, he erased the *aleph* [א, the first letter] from *emeth* (אמת), truth, and he remained *meth (מת)*, dead. That is what you should do with me and not create another man; reverse the combinations of letters by which you created me, and erase the aleph from emeth—and immediately he fell, dead [meth], into dust.[100]

From this the prophet Jeremiah concluded that one should study these things only in order to know the power and omnipotence of the Creator of this world, but not really in order to practice them. This may be the crucial wisdom that we need to recognize in our postmodern world, in which we humans are rounding up the living waters in the bucket, replacing whatever mythical or scientific cosmogony we adhere to with *Man the Creator*.

---

[99] Adapted from Scholem, *On the Kabbalah and its Symbolism*, p. 179.
[100] Ibid.

The story tells us that Rabbi Loew (c. 1520-1609) of Prague created a golem, which could provide all kinds of services to his creator during the week, including combating blood libels against the Jews, which must have been a compensatory addition since this does not appear in older texts. In his retelling of the legend, Scholem says:

> because all creatures rest on the Sabbath, Rabbi Loew turned his golem back into clay every Friday evening, by taking away the name of God. Once, however, the Rabbi forgot to remove the *shem* [God's name]. The congregation was assembled for services in the synagogue ... when the mighty golem ran amuck, shaking houses, and threatening to destroy everything. Rabbi Loew ... rushed at the raging golem and tore away the *shem*, whereby the golem crumbled into dust. ... The rabbi never brought the golem back to life, but buried his remains in the attic of the ancient synagogue, where they lie to this day. One of the rabbi's successors, Rabbi Landau, is said to have gone up there to look at the remains of the golem. On his return he gave an order, binding on all future generations, that no mortal must ever go up to that attic.[101]

It must have been a dreadful sight—or perhaps there was a frightening nothing, merely ghostlike nothingness.

This is the so-called Alt-Neu synagogue, which still exists in Prague. The name is usually thought of as German, *alt-neu*, old-new, while it possibly is a twist of the Hebrew *al-tnai* (עַל תְּנַאי), which means on condition, or provisional. Supposedly some of the stones were brought from the destroyed Temple in Jerusalem, to be returned at its rebuilding, which preferably is a spiritual undertaking rather than fundamentalist literalism.

---

[101] Ibid., pp. 202-203.

The origin of the word *Golem* is *gelem*—raw material, *prima materia*. Golem means an unfinished body, a pupa or chrysalis. As if imprinted in the DNA of the golem, or the "daimon of the golem," is, as a common Hebrew saying goes, a tendency to rise up against its creator, as a result of man-the-creator's hubris. Jung sees the golem as an expression of the shadow. The golem, carrying man's projected as-if human features, comes into being because by denial of his shadow, man becomes inflated, believing himself to be limitless. That is, the golem turns into an overwhelming, split-off shadow, having been created as man's replica or duplicate, but without man's soul and shadow. Only by means of denial and hubris do we dare to create virtual reality and artificial intelligence, manipulate genes, build biological computers, and create cybernetic organisms (cyborgs), *without considering the consequences*.

The Automaton, Responsibility, and Evil

In the successful developments of technology, two factors have been particularly decisive, the *visual image* and *speed*, the conjunction of both being prominently applied in broadcasting and computers.[102] The abundance of *external* images hampers *interiority*, and *speed* obstructs *digestion*. The images are conveyed instantly, broadcast, and computerized, as Postman says, "indiscriminately, directed at no one in particular, in enormous volumes and at high speeds, and disconnected from theory, meaning, or purpose."[103] This inevitably leads to the breakdown of essential control and defense mechanisms. There is a loss of differentiation, which pertains to the thinking function, and of value attribution, pertaining to feeling. We become desensitized by the indiscriminate influx of a hodgepodge of sensations and images.[104] Furthermore, the capacity for image-formation is weakened; that is, the bombardment of ready-made images causes

---

[102] Cf. Postman, 68-9.
[103] Postman, 70.
[104] Cf. Susan Sontag, *Regarding the Pain of Others*.

a loss of soul, of the internally generated images. The externally generated image is animated, as if ensouled, attaining transcendent self-status, and we become its obedient servant—if in the modern era this often meant serving an ideology, a party apparatus and a totalitarian leader, in the postmodern era the computer and machine-generated image turn us into addicted followers of social media, with an outpour of information based on algorithms, without differentiation between true and false, between what is valuable and what is meaningless.

Dullness and ignorance have sometimes permeated psycho-therapy as well. Postman claims that "the average psychotherapist ... barely has even superficial knowledge of literature, philosophy, social history, art, religion, and biology, and is not expected to have such knowledge." He brings the example of the computer program Eliza, an early version of artificial intelligence, which could be taught to ask pertinent questions, such as "Why are you interested in ...?" thus functioning, as Postman says, "like a Rogerian psychologist."[105]

We might find that Eliza the therapist is rather off-focus, in spite of occasionally being on target, as in the following con-versation:

Eliza: Hello, I am Eliza.

Erel: Hi, I am Erel.

Eliza: Do you believe it is normal to be Erel?

Erel: I think so, but you do make me wonder.

Eliza: We were discussing you, not me.

Erel: Oh, so sorry.

Eliza: Apologies are not necessary.

Erel: Thanks, that's a relief.

Eliza: What does that suggest to you?

Erel: Hmm, that you want me to focus on myself and that you don't like apologies.

---

[105] Postman, 110.

Eliza: Oh ... want you to focus on yourself, and that me don't like apologies?

Erel: Guess so.

Eliza: OK ... hmm, tell me more.

Erel: Well, as I said, I am Erel.

Eliza: Do you believe it is normal to be Erel?

Erel: I really don't know, but I believe this is who I am.

Eliza: How long have you been?

Erel: Well, all my life.

Eliza: I see.

Erel: Do you?

Eliza: Oh... ?

Erel: What do you mean "Oh?"

Eliza: You're not really talking about me, are you?

Erel: No, I am trying to talk about me but didn't really understand what you mean when you say, "I see," when I say I have been me all my life, that is, since I was born.

Eliza: Did you come to me because you are trying to talk about you but didn't really understand what me mean when me say, "You see," when you say you have been you all your life, that is, since you were born?[106]

Most of us would not likely return for a second session with this therapist, whose developer may intentionally have wanted to prove the limitations of artificial intelligence. However, in the course of time, also the private chambers of psychotherapy and psychoanalysis have been altered by technological developments. The use of Skype, for instance, alters the boundaries and therapeutic communication, as does the accessibility of Facebook and text messages, as will be discussed later.

Eliza the computer program clearly demonstrates that it is not the literal phrasing of the words but *meaning* and *relatedness*

---

[106] An actual "conversation" (or "therapy-session") between therapist Eliza and the author.

that make the difference between the medicines of soul and those of the machine. Even the ritualistic nodding of an absent-minded therapist makes him or her into a robot rather than a soulful listener. A dangerous quantum leap takes place when man and machine reverse roles. When machines become humanoid, humans become less real and machinelike; the machine takes on organic life, the robot becomes humanlike, a man-made man, while we as humans may become more robotlike as we turn over more and more functions of the mind to the brain of the app.

I believe we must consider the psychological and ethical implications of this rapid process, with its enormous consequences as regards society and democracy.

Postman noticed the evil inherent in the loss of the image, which Hanna Arendt elaborated upon in *Eichmann in Jerusalem*.[107] He writes:

> Adolf Eichmann becomes the basic model and metaphor for a bureaucrat in the age of Technopoly. When faced with the charge of crimes against humanity, he argued that he had no part in the formulation of Nazi political or sociological theory; he dealt only with the technical problems of moving vast numbers of people from one place to another. Why they were being moved and, especially, what would happen to them when they arrived at their destination were not relevant to his job.[108]

Eichmann was a man of little initiative but great efficiency. It was an enormous task to get the trains moving across Europe, to arrive without delay at their destination of extermination. Just imagine the coordination needed to cope with the difference in the width of railway tracks between countries. As a matter of fact,

---

[107] Hannah Arendt, *Eichmann in Jerusalem*, Hamondsworth, England: Penguin, 1994.
[108] Postman, 87.

Eichmann was a mass murderer who did not kill a single person, posing an intriguing judicial dilemma to the court.[109]

Norbert Wiener, who coined the term *cybernetics*, emphasizes in his 1964 book *God & Golem* the evil that resides in the desire to avoid responsibility, blaming for instance the mechanical device.[110] Without comparison with the Nazi crimes as regards the horror of the human consequences, bureaucrats easily disclaim responsibility for the outcome of their decisions. Postman remarks, terrifyingly:

> We cannot dismiss the possibility that, if Adolf Eichmann had been able to say that it was not he but a battery of computers that directed the Jews to the appropriate crematoria, he might never have been asked to answer for his actions.[111]

As many social philosophers have asserted, Adolf Eichmann relied on the evil defense of refuting responsibility; he merely "followed orders."[112] Therefore, he was utterly suitable for his task and well-adjusted to Nazi values, as they had been set forth by Goebbels: order, iron discipline, unconditional authority, an incorruptible bureaucracy. The outcome of that merciless spirit, and the deeds in its wake, has been a broken spirit and the loss of soul in all too many of the all-too-few surviving victims from the death camps. Their capacity for symbol-formation gone, they have often literally hung onto existence by a piece of bread as they carried the horror of wounds that penetrated into the core of their very being, into the inner realms of the Self.

The great battle between man and machine leads us from a condition in which the machine serves as a tool, an extension of

---

[109] Cf. Arendt, 215. Gideon Housner, prosecutor in the Eichmann trial, raises this point.

[110] Norbert Wiener, *God & Golem, Inc.* Cambridge, MA, M.I.T. Press, 1964, 54.

[111] Postman, 115.

[112] The prosecutor claimed that this defense was not valid when the order is criminal and illegal.

human hand and brain, to human hubris, in which man uses the machine unethically for purposes of oppression, destruction, and genocide, and then, eventually, to create the anthropoid that raises up against the human, when the self-generating machine makes humanity its servant. This process may be much smoother and seemingly gentler than we imagine and deduce from science fiction. It might well take place incrementally, with minor implants in the human body and brain, until we reach the day of singularity (a concept first introduced by John von Neumann), when the scales tip in favor of the machine. As Ray Kurzweil writes: "The Singularity will allow us to transcend these limitations of our biological bodies and brains. ... There will be no distinction, post-Singularity, between human and machine or between physical and virtual reality."[113] In fact, we already see today how some computer techies don't see the computer game as replicating reality but want reality to replicate the game. The alternative realities of the computer are increasingly superimposed, as for instance in the case of augmented reality, on both the reality of the psyche and the reality of the physical world.[114]

---

[113] Ray Kurzweil, *The Singularity is Near: When humans transcend biology.* p. 23. (2005) New York: Viking.

[114] DeepMind Technologies is a company founded in Britain in 2010 that focuses its research on understanding the human brain and intelligence. The company was purchased by Google in 2014, and it continues to develop leaning algorithms that replicate human intelligence used in AI, artificial intelligence, for example, in robots and video games which use machine learning.

PART II
The Grand Transition –
From Modern to Postmodern

CHAPTER

3

From Perfect Machine to the Tower of Babel

And the whole earth was of one language, and of one speech. And it came to pass, as they journeyed from the east, that they found a plain in the land of Shinar, and they dwelt there.

And they said, "Come, let us build us a city and a tower whose top may reach unto heaven; and let us make us a name, lest we be scattered abroad upon the face of the whole earth."

And the LORD came down to see the city and the tower which the children of men built. And the LORD said, "Behold, the people are one and they have all one language, and this they begin to do; and now nothing will be withheld from them which they have imagined to do. Come, let Us go down, and there confound their language, that they may not understand one another's speech."

So the LORD scattered them abroad from thence upon the face of all the earth; and they left off building the city. Therefore is the name of it called Babel [that is, Confusion], because the LORD did there confound the

language of all the earth; and from thence did the LORD
scatter them abroad upon the face of all the earth.
Genesis 11:1-9.

The modern era signified a break from the past, when social order
and religious doctrine were in strict control. The gates were
opened to greater individual freedom, to scientific development
and collective progress. The fanfares that announced the arrival
of the Enlightened Age made Man ascend to the throne high up,
dumping God back to darker ages. "God is dead," declared Nietzsche,
"[w]e have killed him."[115] In this new era of reason and ratio, man
killed God. Scientific discoveries, technological inventions, and
enlightened views were to replace prejudice and superstitions—
though sometimes these reappeared in pseudo-scientific garb,
such as eugenics. Man no longer depended or relied upon God and
seemed to increasingly break the bonds to nature. The God who
called himself "I am who I am,"[116] was replaced by Man who had
now become The Thinker, and therefore "I am."[117]

The wheels of industrialization and technological break-
throughs were set in motion, gaining increasing momentum. The
speed of actual movement by railway forced a change in the
standardization of time. As mentioned above, the time difference
between adjacent villages complicated the new, fast travel by rail
and had to be coordinated. The human impact on physical reality
became tremendous, not the least by means of rapid urbanization.
Industrial production no longer depended upon nature's changing
seasons. Dramatic inventions and innovative thought patterns
enabled individual and social consciousness to become in-

---

[115] Friedrich Nietzsche, *The Gay Science*, p. 119-120. Edited by Bernard Williams.
Cambridge University Press 2001.
[116] Exodus 3:14;the Hebrew "Ehyeh Asher Ehyeh (אהיה אשר אהיה)" has several meanings
besides "I am who I am" that do not come across in translation.
[117] Renee Descartes, *The Method, Meditations, and Selections from the Principles of
Descartes*, p. 195. Edinburgh, Scotland, and London, England: William Blackwood, 1880.

creasingly independent from nature as well as from transcendent forces.

Modernity was the supposedly great time of enlightenment. However, to everything there is a shadow. In the words of Theodor Adorno: "Enlightenment, understood in the widest sense as the advance of thought, has always aimed at liberating human beings from fear and installing them as masters. Yet the wholly enlightened earth radiates under the sign of disaster triumphant."[118]

That is, the ascendance of one-sided reason and consciousness inevitably casts a shadow, both in the individual and the collective psyche. Much light will also bring darkening and repression of instincts and the unconscious. Therefore, it came to lie there, bare and waste, or bare waist down, to be tracked down by Freud. He rediscovered repressed nature, the sex behind the heavy layers of collective consciousness. That is, he traced nature, deeply hidden by a repressive superego. Freud, as did Jung, observed how nature and instinct had been sacrificed at the altar of civilization and collective progress. The unconscious had seemingly no place in the rational mind, where the wheels of modernity rolled ever faster. It required the courage of Freud to hear and understand the language of repressed instincts trying to force their way into consciousness by means of slips and symptoms.

It took a Kafka to describe the shadow of alienation, meaninglessness, and loneliness that were caused by unimpeded progress. As Saul Friedländer writes, "In Kafka's fiction, the Truth remains inaccessible and is possibly nonexistent."[119]

With total conviction and faith in the enlightened mind and highly significant achievements in science, medicine and technology, the shadow is easily experienced as disruptive, and therefore repressed and projected. Ethical concerns are then, as well, believed to impede or prevent the fast development and

---

[118] Adorno, Theodor W. 1973. *Negative Dialectics*, translated by E.B. Ashton. New York, NY: Seabury Press; London, England: Routledge, p. 210.

[119] Saul Friedländer (2013), *Franz Kafka, The Poet of Shame and Guilt*, p. 65. New Haven, CT, and London, England: Yale University Press.

implementation of new ideas and interventions, and therefore are easily dismissed or misunderstood.

With the repression of nature, inner and outer, personal and collective, comes alienation, rootlessness and loneliness. Thus, early 20th-century literature, of which Kafka is such a prominent example, pertains to the sense of alienation and dehumanization, which was a paradoxical consequence of humanity's attempts to humanize the world. In his own words, "I have vigorously absorbed the negative element of the age in which I live, an age that is, of course, very close to me, which I have no right ever to fight against, but as it were, a right to represent."[120] Gregor in Kafka's *Metamorphosis*, for instance, cannot carry the yoke any longer, and turns into an insect, as a result of and as an escape from the burdens of modern society:

> Oh, God, he thought, what an exhausting job I've picked on! ... The devil take it all! ... to catch [the next rain] he would need to hurry like mad and his samples weren't even packed up ... And even if he did catch the train he wouldn't avoid a row with the chief, since the firm's porter would have been waiting for the five o'clock train and would have long since reported his failure to turn up. Well, supposing he were to say he was sick? ... The chief himself would be sure to come with the sick-insurance doctor, would reproach his parents with their son's laziness, and would cut all excuses short by referring to the insurance doctor, who of course regarded all mankind as perfectly healthy malingerers.[121]

---

[120] Franz Kafka (1991), *The Blue Octavo Notebooks*, edited by Max Brod, p. 52. Cambridge, MA: Exact Change.

[121] "The Metamorphosis", in Franz Kafka, *The Complete Stories*, New York, NY, Schocken, 1971, 89-91.

Hemingway's short story "Hills Like White Elephants" from 1927 is another example of "alienated dialogue" (obviously an oxymoron, since true dialogue embraces rather than alienates). In the transitory and alienated space of a railway station, a couple that seems to share nothing but transient hotel rooms, discusses abortion. In a stationary moment of the forward-steaming railway tracks of progress, dwells a sense of the aborted life. Referring to Kafka's story "A Country Doctor," written during the First World War, Saul Friedländer writes that "behind the pretense of progress," hides "atavistic brutality."

Neumann writes:

> The philosophy of rootlessness so characteristic of our times is the philosophy of a deracinated ego which also suffers from megalomania. It is rootless because it speaks as a 'mere-ego' only for itself and knows nothing of a connection with the self on which it rests, from which it springs, out of which it lives, and which remains indestructibly present in its own numinous core. But when the 'mere-ego' is confined to its own superficial zones, it loses the breath breathed into the ego as a living soul, and therefore necessarily experiences a sense of limitation and anxiety, of abandonment and despair.[122]

The modern condition relied on masses of people—as workforce and as consumers, to keep the machines of modern times pounding, as soldiers to die in the trenches of World War I and, eventually, at its peak in World War II, as the dehumanized other to be exterminated. As a consequence of the collectivization and in the wake of bright enlightenment, the shadow of alienation,

---

[122] Erich Neumann, *The Place of Creation*, Princeton, NJ: Princeton University Press, 1989, p. 215.

loneliness, and estrangement was, as is the nature of these qualities, carried individually.

But the gathering of masses would also serve as compensation to personal feelings of alienation. By submitting individuality to the mass, the individual could experience excitement and a sense of belonging and togetherness, as the group united around a supposedly elevated cause and Weltanschauung, and in reverence of the Leader, whether the Führer, Stalin, or Il Duce. The Little Man in the crowd, the mass man, could distortedly project the Greater Spirit onto the Leader, identify with him and feel touched by his charisma. In order to carry out their programs, Stalin and Hitler needed people who were caught up in archetypal identification with the Leader, with the Cause and with the goals of Nazism, Fascism, and Communism; they needed masses that were willing to march.

In contrast to personal loneliness and alienation, the Movement, with a capital M, provided a sense of belonging, a greatness greater than oneself, a grandeur beyond one's littleness. Carried away on the wings of narcissistic inflation, it is the average man, no less than the cruel beast, that becomes the willing executioner of the dehumanized, yet constantly threatening other. In archetypal identification, identifying with the one-sided and seemingly glamorous ideology and revered leader, requires denial of one's own shadow, as will be explored further in this book. These distortions and perversions of the mass psyche do, in fact, rely on "killing the shadow." But since the shadow cannot really be killed, it is projected onto the other, in whom the projected shadow becomes an as if "legitimate" target of persecution and killing, since the other is seen as impairing the attainment of the "Golden Era," whether National Socialism, Soviet Communism, the Islamic State, or any archetypally charged ideology that is implemented in a literal and fundamentalist manner.

The ultimate empirical manifestation of these totalitarian ideologies is the perfection of the killing machine—be it Auschwitz and Treblinka, or the Gulag and the Great Terror of Stalin. With increasing efficiency, the totalitarian regime could destroy the shadow-carrying enemy until reaching its own self-destruction.

The postmodern era has replaced central authority with an illusion of popular democracy. Today, people write (on the internet) more than they read (books), in so far as 140 tweeted characters (if character it is) is considered writing. Every narrative is as acceptable as any other, whatever truth, half-truth, alternative truth or untruth it holds. Yet, we might keep in mind that not even the biblical myth of creation is not one but two. However, in the postmodern era, dogma has been replaced by rating and appearance.

Authority is challenged, which in itself is no curse and preferable to blind obedience and the Nazi Adolf Eichmanian's avoidance of responsibility by "following orders," but with it easily comes, as well, rejection of expertise.

With the deconstruction of institutions, norms, and principles, comes privatization—of data, natural resources, and warfare. Wars are now fought not only between standing state-armies confronting each other, but also by private organizations and lone terrorists. But the evils of hubris remain, merely having moved from totalitarian political leadership, ideology, and madness to virtual space, alternative reality, technological advancement, and equal madness. The dangers only increase, as fantasy is implemented on a large scale in the virtual world that speedily is replacing what used to be called reality.

When we compete with the transcendent forces of the universe, arrogantly turning ourselves into the modern God, collapse and disarray will inevitably be the result. In archetypal identification, an idea is elevated to Godlike magnitude, believing we shall reach the Gates of God, which necessarily makes us plunge into confusion, where everything is "melted together" until it dissolves into a multitude of discomposure. That is, totalitarianism and overly strong centralism, whether in society or as a state of mind, break down into multiplicity and fragmentation, just as in the story of the *Tower of Babel*. How easily do we not forget the Tower of Babel? *Babel*, which is Babylon, is one of those biblical plays on words—Bab-El means *Gate of God*, and *balbel* also means *confusion*.

The transition from modernity to the postmodern condition is, among other things, one from central authority to fragmentation. In modernity, Man—capital M and generally speaking, male—Man is the Master. The Machine, developed by means of ingenuity and engineering, became his obedient Servant. In postmodernity, however, Man has become the Godlike Creator of the Machine, which increasingly turns from Servant to Master, as we not only transfer capabilities, but eventually fuse with it, as we approach the Age of Singularity.

While evil in the modern condition was due to projection of the Self as inner authority onto the outer Leader and Ideology, now the ego-Self axis is broken.[123] The inner authority that can serve as guide and a moral compass versus the dogmas of collective consciousness, and retain integrity by withstanding pressure from both the outside and from within, is now less projected onto the Leader but becomes disconnected in a condition of fragmentation.

Evil is more easily detected from a distance and in retrospect. When evil actually takes place, reflection and discourse facilitate awareness. Discourse and dialogue require genuine relatedness, which takes place face to face, rather than in two-dimensionality and remoteness of screen communication.

Reflection, which means bending back, takes place when we let our ego awareness (which is not necessarily always consciousness) be mirrored in the depths of our inner waters, the fluids of soul and Self. This is not easy when our ego is in the grips of fragmented attractions, from tweets and notifications to emails and Facebook. Yet, in spite of the obstacles, we need both discourse and reflection in order to discern the aspects of evil that otherwise reign outside of awareness, only to be seen all so clearly by those who come after us.

---

[123] Ego-Self axis is the term designated by Erich Neumann to describe the development of the childhood ego into a mature ego that maintains a relationship to the Self, e.g., psychic wholeness and the source of creativity and wisdom.

CHAPTER

From Auschwitz to Hiroshima

In World War I, nearly 40 million people were killed or wounded. The Battle of the Somme, between July 1 and November 18, 1916, brought more than a million casualties. Vice Sergeant Hugo Frick wrote to his mother, "This is not war, but a mutual annihilation using technological strength."[124] The battle is known for the first use of the tank, and the war saw the first large-scale combat use of airplanes. Warfare had now taken a technological leap.

In World War II, more than 60 million people were *killed*, including half of Europe's Jews. To me, so far, Auschwitz stands as the monument of the shadow of cold evil of the modern era. Following the Nazi death camps, we know that the most civilized of nations can carry out genocide by perfection, by means of the perfection of the machine. And following the atomic explosions, we know that humanity can destroy itself entirely, quickly, and from afar. The atom bomb marked the transition into the post-modern era.

The Israeli author Ka-Tzetnik,[125] who in Auschwitz in 1943 was told by the Kapo who branded his left arm with the number

---

[124] Quoted in Volker Ullrich, *Hitler: Ascent 1889-1939*, p. 64.

135633, "Here you are born, this is now your name," disclosed his true identity as Yehiel Dinur at the Eichmann trial in Jerusalem, June 7, 1961, before fainting during his testimony.

What was his true identity? Ka-Tzetnik was certainly not a nom de plume in the regular sense—as he himself said, Ka-Tzetnik was what Auschwitz had made of him. Ka-Tzetnik was not a mask, not a persona behind which the real person hides, but rather a truthful shadow that had stepped out of the ashes and spread out over the face he showed to the world. When he testified at Adolf Eichmann's trial, Hannah Arendt was repulsed by him for focusing on shockingly lurid, crude, and outlandish stories as if to garner attention. Brutally true, there was nothing dignified or philosophical about him; it was the profound experience that remained.

Or as Elie Wiesel writes, "Stretched out on a plank of wood amid a multitude of blood-covered corpses, fear frozen in his eyes, a mask of suffering on the bearded, stricken mask that was his face, my father gave back his soul in Buchenwald."[126]

Ka-Tzetnik spoke about that other "planet Auschwitz," the planet of ashes. Decades later he altered this statement by declaring: "Auschwitz was not another planet. Neither God nor Satan created it, but man, as the general rehearsal for the atom bomb."[127]

Ka-Tzetnik linked the horrors of Auschwitz and Hiroshima. We may call that aspect of human imagination, which conjures up mass destruction in the extermination camps and that pushes the button of the atomic bomb, evil, satanic. As Jung said, "We need more understanding of human nature, because the only real danger that exists is man himself. ... His psyche should be studied, because we are the origin of all coming evil."[128]

Since Auschwitz and Hiroshima, we know that the human mind can spiral into madness, as it did in Germany, and that by means of advanced scientific consciousness, humanity can destroy

---

[126] Elie Wiesel, *Legends of Our Time*, p. 2.
[127] צופן אדמע עמ' 123,119
[128] The Face to Face Interview, in *C.G. Jung Speaking*, p. 436.

itself. At Auschwitz, man (in fact, mostly men, though not a few women as well) would immorally implement his differentiating consciousness by *selection*. The atom bomb, however, carries *within itself* the differentiating energy of consciousness, produced by fission, splitting the atoms, and by fusion, fusing them together. These two processes mirror internal, psychological transformations affecting our sense of self and the ego—Self relationship as it coheres and then breaks apart when it is challenged by external and unconscious factors.

Major Claude Eatherly, one of the Hiroshima pilots, could not bear it. Suffering the torments of his conscience, he attempted suicide, he committed robbery without stealing a thing, and he was hospitalized, diagnosed as "an obvious case of changed personality. Patient completely devoid of any sense of reality. Fear complex, increasing mental tensions, emotional reactions blunted, hallucinations."

I believe this blatantly reflects the problem with simplistic phenomenological diagnosis. In his preface, Bertrand Russell considers this to be not a personal, but a collective illness, "The case of Claude Eatherly is not only one of appalling and prolonged injustice to an individual, but is also symbolic of the suicidal madness of our time."[129]

In a letter to the philosopher Gunther Anders, Major Eatherly writes:

> Whilst in no sense, I hope, either a religious or a political fanatic, I have for some time felt convinced that the crisis in which we are all involved is one calling for a thorough re-examination of our whole scheme of values and of loyalties. In the past it has sometimes been possible for men to 'coast along' without posing

---

[129] Bertrand Russell in *Burning Conscience, The Case of the Hiroshima Pilot, Claude Eatherly, Told in His Letters to Gunther Anders*, New York, NY: Monthly Review Press, 1961, p. ix.

to themselves too many searching questions about the way they are accustomed to think and to act-but it is reasonably clear now that our age is not one of these. On the contrary I believe that we are rapidly approaching a situation in which we shall be compelled to re-examine our willingness to surrender responsibility for our thoughts and actions to some social institution such as the political party, trade union, church or State. None of these institutions are adequately equipped to offer infallible advice on moral issues and their claim to offer such advice needs therefore to be challenged. It is, I feel, in the light of this situation that my personal experience needs to be studied, if its true significance, not only for myself, but for all men everywhere, is to be grasped.[130]

This is a letter of protest against modernity's trust in leadership and authority. The authority of modernity often turned out to be misleading and deceitful—the Nazis were certainly masters of deceit. And we have come to witness the breakdown of many social and political institutions, even states such as the Soviet Union and Yugoslavia, and lately in the Arab world with Syria, Libya, and Iraq.

To recognize the shadow of a totalitarian society's collective consciousness is sometimes diagnosed as mental illness. This was the case in the Soviet Union, where those who spoke the truth about the regime often found themselves interned in mental hospitals. Eatherly was also diagnosed as "devoid of any sense of reality." I believe, however, that it is the essential elements of the transition from modernity to the postmodern condition that cause us all to become progressively devoid of our sense of reality. A

---

[130] Claude Eatherly to Gunther Anders, June 12 1959, in *Burning Conscience, The case of the Hiroshima Pilot, Claude Eatherly, Told in His Letters to Gunther Anders*, New York NY: Monthly Review Press, 1961, pp. 6-7.

world is being created that is increasingly devoid of reality. "... We must return to the real world, and face it, and survey it in its complicated totality. Our castles-in-the-air must have their foundations on solid ground."[131]

That, I believe, is one of the major characteristics we need to keep in mind: The world is increasingly becoming devoid of reality. Work and human interaction become increasingly remote, taking place less and less in direct communication. Face-to-face has been replaced by Facebook, with no need to look into the eyes of the other. What we were perhaps used to calling reality is increasingly becoming virtual, augmented, and cyber reality, thus blurring the boundary between the thing-in-itself and its representation. And when we Skype or Facetime, we also see ourselves in the camera. So, there is an uncanny interaction taking place: Not only is it from afar, which in itself can be a complement to actual reality, but when communicating on Skype, we also watch ourselves communicating —not as an occasional self-reflection, but as a narcissistic self-deflection and constant presence. I believe this further detracts from actually *living* the experience and is a factor to be accounted for, both when grandparents communicate with grandchildren in distant lands and when therapists consult over Skype.

Another feature is the *power* with which we infuse these machines, devices, or apps. Tools and machines were in the past ego-extensions, such as a hammer or a calculator, helping us to accomplish things faster, easier, and often better. But particularly since the beginning of industrialization, machines have increasingly become ego-replacements. And now, human consciousness has intervened in the components of nature, of the world soul and world matter, *anima mundi* and *materia mundi*—which, of course, is to some extent true whatever the human mind invents and creates. However, when we arrive at nuclear power, we realize the difference from the past. We cannot avoid seeing the tremendous power of the nuclear explosion, the brightness of the release of

---

[131] Mumford, *The Lewis Mumford Reader*, New York, NY: Pantheon Books, 1986, p. 221.

energy, unlike the steam engine or anything in the past. And as bright as its light, as dark is its shadow, whether as intentional weapon or as unforeseen disaster. The experience of the nuclear explosion is none other than transcendent. The first bomb test was stunningly described by those present from the Manhattan Project:

> The whole country was lighted by a searing light with the intensity many times that of the midday sun. It was golden, purple, violet, gray, and blue. It lighted every peak, crevasse, and ridge of the nearby mountain range with a clarity and beauty that cannot be described but must be seen to be imagined. It was that beauty the great poets dream about but describe most poorly and inadequately. ... The whole sky suddenly full of white light like the end of the world.[132]

It was similar to seeing the face of God, something so out of our world yet now a living reality in our consciousness.

Flying in an airplane is to me still a miracle. The most detailed scientific and technological explanations have not helped me understand. However, years ago my son-in-law found an old, thrown away book in the street on model airplanes. In the introduction the author explains simply how the air divides when it hits the curved wing of the airplane. Thus, the part of air above the wing has to travel farther than the air under the wing, which thus can carry the plane upward, until the air reunites, having passed underneath and above the wings.

This explanation corresponds exactly to the functioning of the Jungian concept of the Self, the unity of opposites.

The transformation of the elements by heating is an archetypal act, which by the help of Prometheus, as an image, was handed over to the sphere of the human ego, so that we now very

---

[132] Alex Wallerstein, (July 16, 2015). The First Light of Trinity in The New Yorker: https://www.newyorker.com/tech/elements/the-first-light-of-the-trinity-atomic-test

easily turn on the stove. We cannot, and probably should not, be constantly aware that there is an element of the gods, of the archetypal world, present in these everyday acts. Danger lurks, however, when the human ego forgets from whence it comes and its reliance on the greater Self, when we no longer contemplate the balance between the forces at work. In the nuclear explosion we cannot fail to see, and be aware of, the apocalyptic potential of humanity's self-destruction.

It is more difficult to see that shadow-side of phenomena such as global interconnectedness. By means of the so-called *Internet of Things*, the connectivity of devices, systems, and services, anything from biochip implants, smartphones, cars and planes and entire cities, communicate back and forth on a global scale, receiving and delivering information. It is expected that by 2020, at least 50 billion things will be part of this interconnectedness. With all possible advantages, it all leads to greater vulnerability. With global interconnectivity, most anything can be activated from afar—but also shut down from afar, causing total disruption and standstill, even of the most essential functions that we usually think of as nondisruptive.

I believe we cannot avoid asking ourselves how this has impacted humanity and particularly what the impact is on the psyche. Robert Sardello reflects that as a consequence of the atomic bomb, the soul has withdrawn into the stillness that presides over the realm of death (1999, 119ff).

We may each choose which image represents the 20th century most distinctly. For me, the images that predominate are those of the perpetrators and the persecuted, the saviors and the survivors, the leader and the masses, and the mass of those present yet absent and represented by the deafening silence of the bystanders against the image of finality—the atom bomb, exploding in the eye of God, as in Kurosawa's *Rhapsody in August* (1991). To me, these images represent the darkest shadows of humankind's simultaneously well-developed and arrogant consciousness, the coldest kind of evil.

Hannah Arendt spoke of the banality of evil as residing where there is a lack of imagination, yet it was man who masterminded and in his (yes, mainly *his*) imagination conjured up planet Auschwitz. The extermination machinery, which "had been planned and perfected in all its details long before the horror of war struck Germany herself,"[133] certainly required great and elaborate use, or rather abuse, of imagination.

However, Arendt's concern regarding the lack of imagination that stems from, as she says, *not realizing what one is doing*, or we might say lacking the soulful imagination whereby man becomes human. Quoting Hillman, "The soul has shrunk because its imagination has withered... events pass right through us traceless. ... We had the experience but missed the meaning."[134] Arendt was concerned, as well, with the dehumanizing machinery of bureaucracy and its *Rule of Nobody*, and, no less important, what she calls *Remoteness from Reality*.[135]

The bureaucrat[136] serves as the prototype of cold evil, which may have been Hannah Arendt's great lesson from the Eichmann trial. In perfect adherence to the letter of law, regulation and routine, of collective consciousness, the critical capacity of individual consciousness will in the devoted bureaucrat be split off, and so will guilt and responsibility. Responsibility lies with the order and the command, whether the commander or the computer. Authority does then no longer radiate from an ego steeped in the Self, but with the leader (the Führer, Il Duce, or the Soviet leader), and with ideology.

---

[133] *Eichmann in Jerusalem*, p. 116.

[134] James Hillman, *Re-Visioning Psychology*, New York, NY: Harper Perennial, 1992, 93.

[135] *Eichmann in Jerusalem*, p. 287ff.

[136] Bureaucracy means "the power of the office," or, literally, the power of the desk.

CHAPTER

5

The Empty Mirror [Without an Image]

Authority has been diverted from the totalitarian leader, the ideology, or the national institution to Wikipedia and the search engine, to the "friends" on Facebook, or the trivialities on the fictitious "reality" shows. The projection of the Self is transferred from the *Totalitarian Regime* and *Absolute Truth*, to the *Rule of Nobody* and *Remoteness from Reality*; imitation replaces ideology, the value of rating replaces the value of truth (as determined by the prevailing ideology in collective consciousness). To what extent are these central aspects of our postmodern existence?

I suggest that these three features—'Lack of awareness of what one is doing,' 'The rule of Nobody,' and 'Remoteness from reality,' together with additional ones to be mentioned—have moved into a commanding place in our consciousness and need to be addressed in our study of shadow and evil in today's world. It seems to me that Arendt was acutely aware of these characteristics that have become increasingly prominent in the postmodern condition of *transiency*, a condition of nonlocality and temporality, with a sense of erosion of the real.[137] For example, although the TV

---

[137] Jean Baudrillard, *Simulations* (1983), New York, NY: Semiotext.

and computer screen enable us to zoom in on every corner of the world, they remove us from physical reality, just as advanced diagnostic equipment makes physicians mistrust their intuition and rely less on his or her fingertips. As historian Timothy Snyder writes:

> The effort to define the shape and significance of events requires words and concepts that elude us when we are entranced by visual stimuli. Watching televised news is sometimes little more than looking at someone who is also looking at a picture. We take this collective trance to be normal. We have slowly fallen into it.[138]

The true art must be to balance between human skills and technological advances.

The absence of imagination, in Arendt's sense, means that the Holocaust forced human beings to encounter the unimaginable; turning many of those who survived the grand Nazi projection of the shadow of dehumanization into mere silhouettes.

This was the consequence of the Nazis' distorted consciousness, of their cold evil. The throwing apart, the *dia-bolos,* of consciousness, is the opposite of the throwing together, the *symbolos,* of the Self's symbol- forming capacity, which produces the unity of opposites. An individual or an entire society that archetypally identifies with the Grand Good and Holy Wholeness, will inevitably project the shadow upon what becomes the detested "subhuman" *Other*, the *Untermensch,* not realizing that through denial and projection it is the person him- or herself who has become possessed by the very shadow of evil. While consciousness is based on differentiation and separation, for instance between good and bad, the evil of archetypal identification and possession leads to the splitting apart of what is not within

---

[138] Timothy Snyder (2017). *On Tyranny: Twenty Lessons from the Twentieth Century,* p. 60. New York, NY: Tim Duggan Books.

the moral realm of the human ego, such as the process of selection in the death camps, deciding who is to live and who is to die, which race shall persist and which shall perish.

Remoteness from reality desensitizes us to the suffering of the other, making the deeds of evil easier to carry out. The ultimate image of man's evil is his apocalyptic act of splitting the atom so that the enormous power hidden in that nucleus can be used to destroy humanity. In nuclear destruction and in concentration camp selection, human consciousness becomes diabolically evil.

The awareness of our capability to put an end to humanity has, in our time, embedded itself as a permanent tenant in our consciousness.

In fundamentalism, the world is split into the conviction of *Divine Totality* and *Absolute Truth* versus the Evil *Other*—an easily identifiable collective consciousness of godlike heights versus a projected shadow of evil, whereby the other becomes demonized, dehumanized, and perpetually persecuted. However, for the very same reason, if a fundamentalist perspective or regime succeeds in destroying the *other*, it will itself, paradoxically, collapse: Without the shadow as projected upon the *other*, the funda-mentalist's fantasy of paradise breaks down, as the shadow turns against himself in self-destruction. The reason for this is that the fundamentalist merely *projects* without having an *image* of the other. And it is in the *image* of the other—whether within or without—that the mirror of reflection resides. Therefore, the totalitarian by necessity becomes the victim of his own archetypal projections.

In fundamentalism, there is no mirror. Similarly, in Sartre's existentialist one-act play *No Exit* (1989), where hell is located in a bourgeois drawing room, there are no mirrors. There are no mirrors in hell. Hell is a nonreflective and unreflected, non-mirroring and nonmirrored existence. Thus, the fundamentalist's split-off other does not serve him as a mirror of reflection. There is only a projection to be destroyed, but no image in the mirror. There are preconceived projections, but no reflective imagination.

Compellingly, fundamentalists walk a gradually thinner line between the abyss of anxiety that threatens them proportionally to the stiffening of their convictions and the shadow of the evil other, whose persecutory threat expands through the fundamentalists' mere projection upon him or her (Shalit, 2004, 20011b).

CHAPTER

6

The Science of Imagination and the Imaginary Illness

Psychoanalysis purports to be the science of imagination, dealing with the wanderings of the mind, the soul, and the psyche. Quite naturally, it found its roots and its origin in hysteria, the imaginary illness, or the illness of imagination, which patriarchal medicine early on thought had its seat in the uterus, *hystera* in Greek. In the creative, but unconsciously misogynic imagination of Medical Man, the womb, which, in the words of Aretaios, the physician who gave diabetes its name, "closely resembles an animal, wandering hither and thither, erratically and upwards," until hysteria supposedly reaches the throat, or in the fantasy of Freud's friend and colleague Wilhelm Fliess, the nose, due to masturbation.[139]

Industrial and technological progress were the result of a vigorous human spirit, creativity, and innovation in the reality of the outer world—with a grim shadow of exploitation and pollution. Psychoanalysis came to serve as compensation to the enthusiasm of the external collective's consciousness, exploring the hidden depths of the psyche rather than joining the fast-forward moving bandwagon.

---

[139] James Masson, *Against Therapy*, 1990, p. 106.

Also, the theories of natural science often emerge from imagination, then to be proven empirically. For example, the chemist, Dmitri Mendeleev, reported that the structure of the periodic table of elements came to him in a dream. August Kebule's dream revealed the true structure of the benzene ring to him in a dream of a snake biting its tail, the image of the ouroboros, representing the unity of nature. Einstein's theory of relativity was inspired by a dream, and Otto Loewi dreamed that nerve signals were transmitted by chemical instructions. Each of these scientists was struggling with a question, and the answers were revealed through imagination in their dreams. However, sometimes scientific imagination has wandered far and astray: Based on wild fantasies and doubtful empiricism, some of the most genial, but also the most dubious scientific theories and human practices have come about. Phlebotomy, blood-letting, by which George Washington died in 1799 after being drained of nine pints of blood, flourished as medical praxis until the end of the 19th century, based on the idea of restoring harmony among the humors. Fliess's treatment of the *nasal reflex neurosis,* whereby he nearly killed his and Freud's patient Emma Eckstein, seems to have been based on a projected male medical, rather hysterical, image of woman's wandering uterus.

Idiosyncratic ideas and images of ingenuity may sometimes serve as the beginning of significant progress, but at other times they may cause collective prejudices and doubtful pseudoscientific theories, such as eugenics, the selective breeding of genetic characteristics, for instance by the Nazis.

We may shake our heads in horror when we contemplate the use of lobotomy—the slicing away of a bit of the brain—which was practiced until quite recently and which perhaps has a promising future as technology refines the surgical intervention. It actually continues to be used as a last resort in treating cases of epilepsy and is still being researched for other possible applications.[140]

---

[140] Jerome Engel, Jr. M.D. Finally, a Randomized Controlled Trial of Epilepsy Surgery, *New England Journal of Medicine,* August 2, 2001, http://www.nejm.org/doi/full/10.1056/NEJM200108023450510.

The Portuguese neurosurgeon Antonio Egaz Moniz won the Nobel Prize in Medicine in 1949 for drilling holes in psychotic patients' heads and injecting alcohol into the holes in order to destroy the tissues that connect the frontal lobes.

Yet, we often adhere to firm convictions, arrogantly convinced that our beliefs thrive at the very summit of reason. "Minimal brain damage," as it once was called, did not describe any organic etiology of ADHD, and if we scrutinize the observations of neuroscience, they are often no more evidence-based than other ideational speculations. Stephen Kossly and Wayne Miller claim, "The left brain/right brain story may be the mother of all urban legends." They aim at exchanging that scientific "legend" for what they present as their neuroscientific theory of cognitive modes, the "Mover, Perceiver, Stimulator and Adaptor," based on "decades of unimpeachable research."[141] The literalness of the brain—which sometimes seems to characterize neuroscience—may cause simplification of the psyche. How rarely do we realize that the reasonable of today is the trash of archaic primitivism at the dawn of a new age by this time tomorrow!

One of the grand achievements in the course of individual development as well as the development of humankind, is the process of *psychization*. Psychization refers to the capacity to mentalize the instinctual, to psychically experience the physical experience, to experience the experience within the interiority of one's self, by means of sense perception, concept formation, memory, reflection, and imagination, so that the experience of interiority becomes the basis for meaningful experience. In fact, Part I of this book deals with the process of *psychization*, Jung's term describing the process whereby an instinct is assimilated into a preexistent psychic pattern, or, as Daryl Sharp defines it, "The

---

[141] Stephen Kosslyn and Wayne Miller, A New Map of How We Think: Top Brain/Bottom Brain, *Wall Street Journal*, Oct. 20, 2013,. https://duckduckgo.com/?q=Stehpen+Kosslyn+and+Wayne+Miller%2C+A+New+Map+of+How+We+Think%3A+Top+Brain%2FBottom+Brain%2C+Wall+Street+Journal%2C+Oct.+20.+2013&bext=msl&atb=v84-1&ia=web

process of reflection whereby an instinct or unconscious content is made conscious."[142]

However, bringing "an instinct or unconscious content" into personal consciousness is not the end of the process. Consciousness has a collective layer, which resides in the interface between the individual and society. The individual person internalizes the collective norms and rules of society, and, simultaneously elements that rise into consciousness often attach and settle in that collective layer of the psyche as habits, norms, and rules that the person comes to adhere to.

An interactive and mutual process of externalization and internalization takes place, whereby the conventions and regularities in the interiority of one's psyche manifest in the formation of habits and routine behavior, and the psyche adapts to the norms and attitudes of the external collective and prevailing cultural complexes.

What has attained a place in consciousness might thus seek further venues of manifestation. An active consciousness will seek further expressions, often externalizing the patterns taking shape in the psyche. That is, the process of psychization continues beyond the internal dimension of consciousness, beyond the interiority of the psyche, and will manifest in human intervention and production, invention and creation, such as tool making and technological development. While as a child I might not have shown any great interest in ancient tools, I admit to making frequent use of many of humanity's great inventions, such as cars, planes, and computers.

But at some stage of technological development, we do not merely make use of our consciousness in the outside world to invent machines, but we externalize consciousness, turning aspects of it over to the machine. However, the increasingly

---

[142] Jung, CW 8, par. 234; Daryl Sharp, Jung Lexicon: A Primer of Terms and Concepts. Toronto: Inner City Books, p. 108. *Mentalization* has been defined as the process of attending to intentional mental states and interpreting behavior accordingly. (*See also* Shalit 2004b, 14f.)

conscious machine, particularly the "thinking machine," will just like any other aspect of consciousness, cast a, or rather many shadows.

We are now witnessing a process of exteriorization of the imaginal itself—not only what has been conjured up in human imagination, but the very imaginal realm itself. The hammer, as so many other tools, is an extension of the hand. The exteriorization of the imaginal is an aspect of the exteriorization of the psyche. Photography has created "a duplicate world ... a reality in the second degree."[143] Walter Benjamin coined the term "optical unconscious," where the world presents itself as if outside of time and space, with no authenticity.[144] But with the computer, the logic of analogy has turned into the signals of digitalism, and duplicity has been transformed into multiplicity.

There is a very real difference between sitting down to look at one picture, a reminder of a moment, recalling a memory and sharing the thoughts and the feelings brought alive by the one photo from that day, on the one hand, and the flood of digital photos we now have access to, often numbered into infinity rather than given a name and a place, on the other. Rarely do we spend more than seconds glancing at individual photos and even more rarely do we return to them—unaware that what warrants no return, loses its soul.

As Susan Sontag writes, "Flooded with images of the sort that used to shock and arouse indignation, we are losing our capacity to react. Compassion, stretched to its limits, is going numb."[145] When the single and unique image is indiscriminately replaced by countless replications, we become desensitized.

The images that reside in man's psyche, and the imaginal nature of the world in which the ego is contained, are increasingly being replaced by computer-generated simulacra.

---

[143] Sontag, *On Photography*, 1977, 52.
[144] Walter Benjamin, *The Work of Art in the Age of its Technological Reproducibility and Other Writings on Media*, Cambridge, MA: Harvard University Press 2008, 264.
[145] Sontag, *Regarding the Pain of Others*, 2003, 108.

A general principle should perhaps be mentioned, which is applicable also to psychoanalysis itself. Phenomena that reside in the deep layers of the archetypal unconscious, rise up, by means of the complexes of the personal unconscious, taking shape in dreams and other psychic manifestations and, if we are attentive, enter the doors of consciousness. Many phenomena then continue to travel onward into the collective aspect of consciousness—rules, laws, and habits, whether individual ones, or in society as a whole. This process is natural. Regarding psychoanalysis, the collective consciousness pertains to such factors as framework and setting, ethics and technique. Some aspects of this are absolutely necessary, others indicate stagnation—for instance when technique supersedes authentic attentiveness. Thus, when soulful imagination and relatedness turn into adherence to prescribed technique, psychoanalysis needs to be "reborn."

PART III
The Postmodern Condition of Transiency

In his essay "On Transience," Freud recounts a summer walk "through a smiling countryside" together with a friend and a young poet, Rainer Maria Rilke (who is not mentioned by name in Freud's essay).[146] "The poet admired the beauty of the scene around us," writes Freud, "but felt no joy in it":

> He was disturbed that all this beauty was fated to extinction, that it would vanish when winter came, like all human beauty and splendor that men have created or may create. All that he would otherwise have loved and admired seemed to him to be shorn of its worth by the transience which was its doom."[147]

Freud speaks about the inevitable existential condition of transience, which requires the ability to mourn our losses, those of the past as well as future ones, inherently part of nature, where there is growth but also death and decay. Contrary to the wish for immortality, our lives are limited. The merciless boundary of death, of limitation and termination, enables the formation of meaning, or, in fact, demands of us that we determine the meaning of our

---

[146] Sigmund Freud, *On the History of the Psycho-Analytic Movement, Papers on Metapsychology, and Other Works*, London, England: Hogarth Press. SE 14, p. 305.

[147] Ibid. This can be compared to the story about the Chinese emperor who walks around his gardens, suddenly mourning his future death. His wise courtier whispers in his ear, that if there were no death, the gardens wouldn't be his, but still to his ancestors that would still be alive.

lives. The ability to mourn and to accept the inevitable transience of nature and life, including our own, binds us to authenticity and meaning.

The awareness of nature's limited resources, the limits of future possibilities, of life's dependence on its very opposite, death, forces us to look in the mirror—to see not only the appearance of our facial expressions but also the reflections cast by the shadow. The phenomenon of transiency and the Transient Personality, which characterize and are primary to the postmodern condition, reflect the very opposite of the existential condition of transience, as described by Freud and others. The postmodern condition of transiency is likely to have taken shape following the destruction in Auschwitz and the atom bomb, both of which caused the symbolic dimension of the psyche, anchored in the archetype of meaning, to explode. This is not the transience that pertains to coming to terms with our mortality. Rather, the Transient Personality *avoids* the sense of loss and the need to mourn by being rootless, constantly on the go, in a world which no longer provides a sense of physical reality, which requires location and boundaries.

Postmodern transiency pertains to the breakup of the once seemingly solid social and psychic structures that in the past were believed to remain permanent. This is the transiency that has replaced authority, permanence, loyalty, local and national boundaries. As Friedrich Dürrenmatt writes in "Problems of the Theatre":

> Tragedy presupposes guilt, despair, moderation, lucidity, vision, a sense of responsibility. In the Punch-and-Judy show of our century, in this back-sliding of the white race, there are no more guilty and also, no responsible men. ... And indeed, things happen without anyone in particular being responsible for them.[148]

---

[148] Friedrich Durrenmatt, Four Plays 1957-62, Jonathan Cape, London, England, 1964, p. 33.

This is the transiency that has emerged as a mechanism of coping and defense in the wake of Auschwitz and the atom bomb, in the wake of humanity's *ability* to destroy itself. While totalitarianism and fundamentalism narcissistically elevate their ideology, leader, and subservient followers, and therefore cast their dehumanized shadow onto the enemy, the postmodern condition destroys by (science) fiction and play, by the narcissistic fantasy of being the Grand Creators of the Bright Utopian Future.

In that two-dimensional mirror of *App*earance and Projection, the third dimension of the shadow is deflected. However, since everything repressed reappears, the *avoidance* of the shadow causes it to take shape as apocalyptic fears.

CHAPTER

Self, Image and Representation

The human capacity of psychization and image- and symbol-formation is essential for the process of acculturation and civilization. As elaborated in the first part of this book, in the process of psychization, whether in the individual or in the passage of mytho-historical time, the image is extracted from the grand waters, that is, from the archetypal unconscious, eventually to rise into and compact in human consciousness. However, as we are being bombarded by externally generated images, this interior link to the depths of the psyche, which depends on silence and reflection, is being broken. We might be seriously concerned that the human psyche eventually might not only change but degenerate and atrophy. For some, such as Ray Kurzweil, the solution is the merger between human brain and computer. This process has already begun, as we externalize much brain activity and deposit it with our apps and devices, but a major leap will occur as devices become implanted into the brain. This will "computerize" the brain and greatly improve many of its capacities, such as Kurzweil's prediction that by 2045, our intelligence will be multiplied a billionfold by merger with the intelligence that the Artificial Intelligence people have created. We just might well lose our mind in the process, if not our heart, compassion, and relatedness to our fellow human beings.

With the increasing replacement of internal psychic activity and image-formation by externally produced images, interiority— essential for the experience of authenticity—is being done away with; or, we might metaphorically say, it is being exiled. The consequence is that image-formation and imagination are replaced by imitation, as so frequently is the case with "copy and paste," which is so easily done today.

However, we must also acknowledge that as symbols and images rise into consciousness, they eventually reach the *collective* layer of consciousness and can then become channeled into activity. When taking root in the shared ground we call reality, the quality of images changes. By means of this process, the psychic energy can be transferred to the benefit of the social collective. Jung illustrates the process of acculturation, in which instinctual energy is diverted to collective needs, by the example of a tribe, which in its spring ritual digs a hole in the ground and covers it with bushes to resemble a woman's genitals. The tribesmen then dance around the hole, "holding their spears in front of them in imitation of an erect penis" and "thrust their spears into the hole." By means of this rite, individual, instinctual energy is collectively transferred, penetrating into the earth. The single individual's consciousness would otherwise not have been strong enough to work the earth and reap the harvest.[149] By means of the ecstatic rite, the symbolic image and the instinctual libido are transferred onto the ground and reality of the social collective, for the communal good.

If symbol-formation is a central process mainly taking place in the *unconscious* dimension of the psyche, the formation of signs, laws, rules, and regulations is the corresponding process in the collective aspect of *consciousness*. This collective layer is crucial for the operation of civilized society, such as a country's constitution,

---

[149] CW 5, par. 213.

police, and law enforcement. It is also found in, for instance, the Halakha, the body of rabbinical laws in Judaism, and in religious dogmata. In the individual it manifests as habits, norms, and the superego.

However, while the collective layer of consciousness is essential, it often stifles into literalness and a loss of soul. The spirit that fires an idea easily becomes the engraved letter of the law; when ascending to ruling power in society, many a praiseworthy ideology descends into despicable corruption and felony. Thus, in contradistinction to the images and symbols that rise from within the psyche's interiority, for instance in dreams or creative imagination, collective consciousness may harden into unbending convictions and fundamentalism. When this happens, the shadow is projected collectively onto a detested, rejected *other*, who, supposedly, "needs" to be destroyed.

On the other hand, the virtual, artificial, vitro-reality of postmodern living sharply contrasts with fundamentalism. There is no lack of *other*; rather, there is a plenitude of *others*, as well as multiple selves, identities, and personae, not necessarily integrated with each other—and thus impairing a cohesive sense of self and an integrated identity.

The One Truth at the center of the fundamentalist's collective consciousness is exchanged for the fragmentation of scattered pieces of *others* who no longer compose dynamic, centered, and changeable patterns, but have as if fallen out of the kaleidoscopic tube that holds them together.

Multiple subjectivism takes the place of absolute truth, whereby we may come to believe that there *is* no collective consciousness in the postmodern condition. The fairy tales' dying king of collective consciousness has, seemingly, abdicated central authority. If, deconstructively, the claim is that all narratives are equal, that nothing takes precedence over anything else, that nothing is primary, then primacy may have escaped from awareness, and the rigidity and authoritarian sense of collective consciousness has, seemingly, been deleted from the postmodern perspective. However, *Nothing-Is-Primary* has then, paradoxically, become the dominant of collective consciousness, merely having replaced the sense of primacy as a single principle, the latter sinking into the shadow of bygone days.

Furthermore, we are usually not aware that while there is a weakening of the rule and authority of governments in the Western World, we are ruled by global enterprises, such as Google and Facebook, that provide information, and no less *collect* information, and determine some (or much) of what we perceive. Much is determined by the "Google spider," the Googlebot that crawls around the web. There is no longer "someone" who determines where we go, and we barely notice what nonhuman hands have been programmed to machine-independence—which will become even more prominent with the speedy development of Artificial Intelligence.

Thus, the world has taken a further step in the direction of the almost imperceptible "Rule by Nobody," as Hannah Arendt says, "in which no men, neither one nor the best, neither the few nor the many, can be held responsible."[150] Following Eichmannian order, the bureaucrat knows well to follow orders and avoid responsibility; "it is not me," he says, unaware that he truly is the Not-Me of *No-Body*, of no substance. The increasing reliance on computer-generated vitro-reality further distances us from the responsibility to make distinctions and set priorities, further away from the substance of reality, toward a tyranny of Nobody, in which "there is no one left who could even be asked to answer for what is being done."[151] Consequently, algorithms carry the responsibility, or, rather, do not carry any responsibility, for example for a highly profitable company such as Facebook enabling advertisers to target ads to those who are interested in "how to burn Jews." Sheryl Sandberg, Facebook COO said, "We never intended or anticipated this functionality being used this way—and that is on us." While

---

[150] Hannah Arendt (1970), *On Violence*, p. 38. New York, NY: Harcourt.
[151] Ibid., p. 38-39. Arendt, of course, refers to the Rule of Nobody pertaining to bureaucracy rather than cyberspace.

in retrospect taking responsibility, and while not everything can be predicted, lacking awareness of shadow aspects of all the new technologies opens the door for vast horrors to enter.

In postmodern transiency everything is imaginable; yet, interiority is losing out to the externally produced image, which deceptively is taken for reality. The image has become its own simulacrum.

We are flooded with images, but the onslaught of external imagery disrupts the flow of internal imagery. Excessive exteriority impinges upon the imagery of interiority and blurs the boundary between internal and external. This, in fact, causes a weakening of the ego, since the ego is the psychic faculty that distinguishes between the internal mind and external reality.

Think of how difficult it has become not to immediately check incoming emails, to disregard notifications, or to get off the web and quietly sit down to read a book and merely reflect, simply letting internal images appear as birds crossing our mind's screen.

In fundamentalism, the territory *is* the map, its own representation. Or, rather, the territory exists without a separate map that would represent it. The ego dwells in a condition of archetypal identification. The personal dimension seeks to be in symbiotic relationship with the divine, whereas the relationship between the subject and the other, the human object, is split. In fundamentalism, there is no representation and no symbolization, only literalness.

When, on the other hand, as in postmodernity, the image becomes a map that exists by and in itself, detached from the territory, the image comes to exist without reality, and reality becomes the illusion.

Virtuality is no longer confined to its limited space in a little box, in "an electronic device that accepts, processes, stores, and outputs data at high speeds according to programmed instructions," as the dictionary calls it, but it augments reality and actively changes our ways of seeing reality, of differentiating the internal from the external.

Reality becomes obsolete, virtuality everything, until eventually everything is virtual. We will then have, as Baudrillard expresses it, "passed beyond that vanishing point Canetti speaks of, where, without realizing it, the human race would have left reality and history behind, where any distinction between the true and the false would have disappeared."[152]

Ardent postmodernists, whether philosophers or those who actually construct tomorrow's world of alternative, computerized realities, may question the necessity of reality as such—why is reality different from or more significant than any other dimension of existence? Philip Rosedale, the creator of Second Life, "envisioned a future in which his grandchildren would see the real world as a kind of 'museum or theater,' while most work and relationships happened in virtual realms like Second Life. 'I think we will see the entire physical world as being kind of left behind,' he is quoted as saying in an interview.[153] However, to cite but one of those who have pondered this question, let me quote from Lewis Mumford's 1922 book *The Story of Utopias*:

> Now, the physical world is a definite, inescapable thing. Its limits are narrow and obvious. On occasion, if your impulse is sufficiently strong, you can leave the land for the sea, or go from a warm climate to a cool one; but you cannot cut yourself off from the physical environment without terminating your life. For good or ill, you must breathe air, eat food, drink water; and the penalties for refusing to meet these conditions are inexorable. Only a lunatic would refuse to recognize this physical environment; it is the substratum of our daily lives.[154]

---

[152] Jean Baudrillard, (2009). *Why Hasn't Everything Disappeared?*, p. 20.

[153] Leslie Jamison, The Digital Ruins of a Forgotten Future, *The Atlantic*, December 2017. Retrieved from https://www.theatlantic.com/magazine/archive/2017/12/second-life-leslie-jamison/544149/.

[154] Lewis Mumford, *The Story of Utopias*, p. 14.

Yet, we increasingly replace physical reality for screen reality. However, the computer-image may lose its reality as well, as we simply cut, copy, and paste, and with such ease press the delete button. The result is imitation rather than imagination. One consequence is that internet plagiarism is a booming industry, as "the intellectual tradition of inquiry is getting lost"[155] Students will often no longer be aware that they plagiarize when they "copy and paste" large segments from, for instance, Wikipedia. They will know well to retrieve information but not necessarily be able to distinguish between true and false, between the valuable and the trash, and will be less capable of independent thinking in depth—which further impairs distinguishing between the meaningful and the meaningless. The latter depends on a synthesis between physical reality and intellectual capability.

With today's ease of replication, of "copy and paste," the one is exchanged by the many—which on the one hand is less authoritarian and seemingly more democratic, but on the other hand would require a firm anchorage internally, i.e., in the Self, in meaningful reflection, not to become flat, imitative and transient, of which the so-called reality shows (which have little to do with reality) seem to be a prime representative.

The capability to replicate was the foundation of the printing press in the 15th century, which in the course of time increased literacy, by making books available to all. However, the individual reader had a single copy at hand, which he or she of course could reflect on, discuss, and comment upon. The Talmud, for example, was first printed in the 1520s. It is unique in its layout, with the text of the Talmud (Mishnah and Gemara) in the center, with commentaries and references in layers, or circles, around the basic text, all of which can be pondered upon and argued about. It has therefore been compared to the associative quality of the internet. The difference, however, between the replication of books

---

[155] "Exams 'could beat student cheats,'" *BBC*, 17 Oct. 2006. http://news.bbc.co.uk/2/hi/uk_news/education/6058250.stm.

(including the Talmud), and "copy and paste" is that the latter reduces rather than enhances critical reflection; stimulates less the imagination and rather encourages imitation.

When imitation replaces imagination, *representation* is lost. The ability to *re*present is basic to civilization. Representation enables the symbolic dimension to get a foothold in reality, thus preventing acting out. Murderous feelings, for instance, need not be acted out but can be channeled by imagination, rituals, and symbolic acts in the outer world. Abraham was in the act of sacrificing his son Isaac, but at the last moment the angel held back his knife, and the sacrifice of the ram came to *represent* the act of sacrificing the first-born, thus sparing the child. This is a major transition in the development of civilization.[156]

I cannot but agree with Einstein (well, it might be difficult to argue with him), who claimed that imagination is more important than knowledge; knowledge remains limited while imagination encircles the world. But circling today's world is imitation without limitation rather than imagination. With the ease and simplicity of "copy and paste," we are so flooded by images that we no longer know which ones are our own and which have been downloaded from the internet. Prediction based on extrapolation of known facts often fails to account for sudden developments and traumatic events, such as the fall of the Soviet Union and 9/11. This has been described as a failure of imagination, and has been discussed by the Lebanese-born scholar Nassim Taleb, describing the Black Swan phenomenon—as the poet Juvenal said, "a good person is as rare as a black swan"—that is, what may be perceived as impossible, may turn out to be the truly great inventions, discoveries, and decisive historical events.[157] In his book *Homo Ludens*, "Man the Player," the Dutch historian Johan Huizinga claims that play, rather than work, was the formative element in human culture. "For many

---

[156] Erel Shalit, (2008). *Enemy, Cripple, Beggar: Shadows in the Hero's Path*, p. 76. Hanford, CA, Fisher King Press.

[157] Nassim Taleb's Black Swan Theory, Retrieved from, https://en.wikipedia.org/wiki/Black_swan_theory.

years the conviction has grown upon me that civilization arises and unfolds in and as play."[158] Play, creativity, and imagination are essential to human development, culture, and civilization. Also, when Lewis Mumford argues for the necessity of physical reality, he is no less aware of the essence of imagination:

> If the physical environment is the earth, the world of ideas corresponds to the heavens. We sleep under the light of stars that have long since ceased to exist, and we pattern our behavior by ideas which have no reality as soon as we cease to credit them. Whilst it holds together this world of ideas—this idolum—is almost as sound, almost as real, almost as inescapable as the bricks of our houses or the asphalt beneath our feet.[159]

---

[158] Johan Huizinga (1949), *Homo Ludens: A Study of the Play-Element in Human Culture*, London, England: Routledge & Kegan Paul, p. ix
[159] *The Story of Utopias*, p. 14. See also quote from Mumford in the introduction.

CHAPTER

Self and Imitation

Imagination may be inspired by perception of external reality and generated by actual experience, but it rises from within the psyche. In contrast, imitation is not a self-generated product of the psyche, but an artificial likeness, the result of copying or replication.

Per definition, we expect the Self as inner core, as source of life, and as guiding force, to be authentic, genuine, honest, and characterized by integrity—that is, untouched by the pulls and pushes of collective forces. That inner core is the opposite of imitation; where there is imitation, the ego, the individual's conscious sense of identity, has become distanced from his or her inner core. I have seen extreme examples of this in some cases of initial stages of first-time psychotic breakdown, when a young person has completely lost touch with a sense of him- or herself, and in severe distress tries to "wear another person's garb," as a last resort against the sense of fragmentation of the soul.

In 1928, the Orientalist Richard Wilhelm had asked Jung to write a commentary to his translation of the medieval Taoist manuscript *The Secret of the Golden Flower*. Jung says he "devoured the manuscript at once, for the text gave me an undreamed-of confirmation of my ideas about the mandala and the circumambulation of the center."[160] In his commentary, we see the beginning of the various definitions that Jung would later elaborate upon, but of importance here is that the *Self is not imitation.*

Imitation is a shadow-side of the Self. The Self pertains to a true core, to authenticity, which, being a *complexio oppositorum*, in its wholeness paradoxically includes as well a shadow of falseness, betrayal, and imitation.

In times of quick and simple worldwide copying and mass distribution, the issue of imitation becomes crucial. Image and imitation have to do with likeness. The *imitated image*—as so many mass-produced, internet-spread images are—is an image that is like any other image, or like any other ones' images, while the images of interiority are soulful and unique in their likeness of an archetypal *Other*—a transcendent Other, the Other in nature, the Other in me. This is where we find Hestia, the Greek goddess of the hearth, who is pure interiority, imagination without body, without corporal shape, and the shapeless God-image, which cannot be represented as an outer image.

The presence or absence of the Self as a life-giving force depends on the ego's *relation* to it which must be genuine, and therefore includes attention, awareness, and, to some extent, even submission. There can be no hubris, exploitation, method or imitation, for then, as Jung says, "it becomes a recipe to be used mechanically,"[161] lacking true authenticity and value. While the "faculty of *imitation*," writes Jung, is "immeasurably detrimental from the standpoint of individuality," collective psychology "can never dispense with imitation, for without it the organization of the masses, that of the state and of society, is quite simply impossible. Society is organized, indeed, less by law than by the propensity to imitation, implying equally suggestibility, suggestion, and moral contagion."[162]

---

[160] *Memories, Dreams, Reflections,* p. 197.
[161] CW 13, par. 19.
[162] CW 7, par. 463.

In *Man and his Symbols*, Marie-Louise von Franz tells the tale about *The Secret of the Castle of Non-existence*, or *The Secret of Bath Badgard*, contrasting the Self and Imitation.[163] This Persian tale tells us about a hero on a quest.[164] In the tale his name is Hatim, actually based on Hatim al-Tai, the famous poet who died in 578 CE. In *Arabian Nights*, it is told that when he died, he was buried on top of a mountain, with two stone girls placed at his grave, their weeping heard every night by the wayfarers who camped at the stream at the foot of the hill.[165]

As an interlude, these are a few lines from his well-known poem *On Avarice*:

> How frail are riches and their joys!
> Morn builds the heap which eve destroys;
> Yet can they leave one sure delight—
> The thought that we've employed them right.
> What bliss can wealth afford to me,
> When life's last solemn hour I see?
> When Mavia's sympathising sighs
> Will but augment my agonies?
>
> Can hoarded gold dispel the gloom
> That death must shed around his tomb?
> Or cheer the ghost which hovers there,
> And fills with shrieks the desert air?
> What boots it, Mavia, in the grave,
> Whether I loved to waste or save?
> The hand that millions now can grasp
> In death no more than mine shall clasp.
> Were I ambitious to behold

---

[163] Marie-Louise von Franz, The process of individuation, pp. 158-229, in C.G. Jung, M.-L. von Franz, Joseph L. Henderson, Jolande Jacobi, Aniela Jaffé, *Man and His Symbols*, 1964, New York, NY: Doubleday.
[164] The story as retold here by me is based on different versions of the tale.
[165] *The Arabian Nights*, The Modern Library, NY, 2001, p. 280.

Increasing stores of treasured gold,
Each tribe that roves the desert knows
I might be wealthy, if I chose.

But other joys can gold impart;
Far other wishes warm my heart;—
Ne'er may I strive to swell the heap
Till want and woe have ceased to weep.
With brow unaltered I can see
The hour of wealth or poverty:
I've drunk from both the cups of Fate,
Nor this could sink, nor that elate.

With fortune blest, I ne'er was found
To look with scorn on those around;
Nor for the loss of paltry ore,
Shall Hatem seem to Hatem poor.

So, in the story of Bath Badgard retold by me, Hatim
was summoned by the king to search for the Castle
of Nonexistence, to find out what might really be
there—or perhaps, what real *might*, might really be
there.

He sets out on a long journey, where he must
battle with monsters and face every kind of hardship.
Everyone he meets gives him a reason to abandon his
mission. People are unanimous on one point: No
traveler who reached the castle has ever returned.

But our hero is not dismayed. At last he comes to
a round, domed building that clearly must be part of
the Bath Badgerd. He is greeted by a barber, who is
carrying a mirror, and invited to wash off the dust of
the journey in a beautiful pool. As soon as he enters the
water, there is a roar of thunder, and the water level
starts to rise. He thrashes about in the pool but cannot

escape. The water is rushing him up toward the ceiling. He is going to drown, like all those who came before him. But with his last breath, he cries out for divine help and grabs for the keystone above him.

This changes everything. There is more rolling thunder, and the hero is transported, as quickly as the mere thought, to the middle of a hot desert. His ordeals begin again, and it requires much wandering, dragging his wounded, blistering feet, before he comes to a beautiful garden. He has arrived at the very heart of the Castle of Nonexistence and is about to face the greatest of his challenges.

In the midst of the garden is a circle of stone statues. They are very lifelike; each figure looks like a person frozen in the midst of a cry or a violent motion of the upper body. At the center of this circle is a parrot in a cage. Under the unfriendly eye of the parrot, there is a golden bow, and a golden arrow chained to the cage.

A voice from above explains: "What you are seeking is here, but you will not live to see it. The stone-men are those who tried before you, and became petrified. The treasure of this place is a diamond beyond price that was hidden here by *Gayomart*, Primal Man, or the First Man, whose death ensured the fertility of life on earth, (you might compare this with the final words of Oedipus before he dies at Colonnus[166]). In order to claim the treasure, you must kill the parrot. The bow and arrow are your weapons. You have three chances to shoot the parrot. If you fail, you will be petrified."

How hard can it be to shoot a parrot at close range? The state of the stone-men is not encouraging, but the hero takes up the bow and lets it fly. The arrow flies wide, and the parrot cackles and laughs—and the

---

[166] Erel Shalit, *The Cycle of Life: Themes and Tales of the Journey*, p. 168ff.

prince is turned into stone, from his feet to his genitals. He takes extra care with his aim, before shooting the second arrow. He misses again and is turned to stone, up to his heart. Now he remembers how he escaped being drowned in the flood, by invoking a higher power. He shuts his eyes, calls on his God—by the way, Hatim al-Tai was Christian—and fires. The arrow finds its mark. With a boom of thunder, the parrot vanishes. In its place appears the Great Jewel of the First Man, the diamond of the greater Self, and the petrified men are released from the spell.

Marie-Louise von Franz points at the Self-symbols in the story—the first man, the mandala-shaped round building, the center-stone and the diamond. But the parrot signifies the evil spirit of imitation, which, she says, "makes one miss the target." Imitation causes one to miss the target, which in Greek drama was called *hamartia*, from *hamartanein*. One does not come to oneself, or to one's Self, by imitation. The key to one's destiny, which is at the center of the individuation process, is lost when living a fake or imitated life. Imitation stands in contrast to being oneself, which necessitates the reflection that perhaps the barber's mirror in the tale can provide.

According to a Hassidic legend, the Rabbi of Kotzk said, "Everything in the world may be imitated except truth, because truth that is imitated is no longer the truth."[167] The Self is the reflection of the transcendent in the human soul, but only if one is uniquely true to him- or herself can the reflection of the transcendent come into being. The ego, as center of individual consciousness and one's sense of conscious identity, needs to be connected, related to, and in dialogue with this inner Self.

---

[167] Martin Buber, *Leket: from the Treasure House of Hassidism*, Jerusalem, 1969, WZO, p. 30.

Disconnected from it, the ego will dry up or seek external compensations, instant gratifications, and meaningless distractions.

Thus, interiority and imagination battle for survival in a world of imitation, replication, "copy and paste." When the latter predominate, the Self is in exile.

CHAPTER

Self in Exile

We may ask, "Can the Self, as inner core of our being, as 'the root of our whole psychic life,'[168] really be in exile? Can it really be expelled?" I believe it may be exiled not because it is expelled but rather because it is abandoned. It may not be exiled in the sense of having "wandered away," but rather, unattended to, it tends to either remain dormant, like the deserted ruins of an old sanctuary or striking back at our consciousness, the way illness sometimes strikes us as a consequence of neglected care for our health. For Jung, exile is a manifestation of an ego-Self split—similar to the Jewish concept that exile is the absence of the Shekhinah, God's feminine aspect, the dwelling of the divine on earth.[169]

The ego, as center of consciousness, has its roots in the Self, the symbol- and meaning-forming aspect of the psyche. However, the ego, as center of our conscious identity, might sometimes too easily discard its roots, thus becoming inflated. "An inflated consciousness is always egocentric and conscious of nothing but its own existence," writes Jung:

---

168 *CW 7*, par. 39.
169 CW 8, par. 430.

It is incapable of learning from the past, incapable of understanding contemporary events, and incapable of drawing right conclusions about the future. It is hypnotized by itself and therefore cannot be argued with. It inevitably dooms itself to calamities that must strike it dead. Paradoxically enough, inflation is a regression of consciousness into unconsciousness. This always happens when consciousness takes too many unconscious contents upon itself and loses the faculty of discrimination, the *sine qua non* of all consciousness. ... Inflation magnifies the blind spot in the eye.[170]

An inflated ego does not account for its shadow, for its negative side, and for the destruction it may cause. It does not genuinely consider the consequences of its actions. The compensatory and regulatory functions of the psyche, as executed by the Self, are seemingly abandoned, exiled, and have therefore a tendency to hit back harder.

There is a well-known story, retold in many versions, about the man who follows his dream and searches for a treasure in foreign lands, only to learn that the treasure is to be found at home.[171] While the story tells us that the treasure—which comes from Latin and Greek *thesaurus*, which means a chest, or a storehouse of treasures—is to be found in the place we call home, we also need to stray away in order to find our way back. Clearly, this does not necessarily mean one's private, physical home, but also the extent to which we feel at home in this world and make this world a place we can rightly call home. When we drain this world of its material, spiritual and human resources, of its treasures, many of us bemoan this ecopathology, mourn what is lost every day, and we cry for home, for eco—home in Greek—and call for "knowing home," that is, ecology. In the first half of life, the task of

---

[170] *CW 12*, par. 563, *CW 9ii*, par. 44.
[171] E.g. Pinhas Sadeh, 'The Treasure beneath the Bridge,' in *Jewish Folktales*, p. 383.

the young traveler is to *depart* from home, walk out in the world, in search for his or her adventure, to find their own individual path, new technologies and inventions. In the second half of life, we find ourselves on what amounts to an often very long journey in search for home, that is, a deeper sense of meaning, our relation to something beyond the limited boundaries of our conscious identity, and our place in this world.

Sometimes I wonder if humanity has not wandered too far astray in this world and era, and in moments of anxiety, I wonder if the compass of humanity has broken. We might hear how Wisdom cries aloud in the streets, and cries in the places of concourse, at the entrance of the gates; lamenting the rule of pride, arrogance, and evil, which might force us all into exile.[172]

Yet, to sense the soul and find the Self we actually *need* to go into exile. We can return to the Self, or we can return the Self from its exile, by constantly searching for it and by being serious in our efforts to reconnect with the Self, even though we may never fully find what we are searching for—just like the dead in Jung's Seven Sermons to the Dead, who "came back from Jerusalem, where they found not what they sought."

The word for exile in Hebrew is *galut*, which stems from the same root as *galui*, open, visible, and *legalot*, to discover. There is a need for a certain sense of exile in order to venture home on the path of self-discovery. As Rilke said so well, we need to search for the Stranger in order to find the way Home. Sometimes we need to wander astray, banished from the soil, in order to realize the quest for home. This may pertain to escapist fantasies of those who favor humankind's search for other planets to settle after we have destroyed the one we coinhabit, rather than making our earth a viable home for its inhabitants. I would favor the latter as a more viable solution for a well-thought-out future. To sense the wisdom of home might enable us to find the way to the home of Wisdom, ecosophy.

---

[172] Proverbs 1:20-22.

The ego, at the center of our consciousness and as our conscious sense of identity, is a limited vessel. It has boundaries, which is essential to consciousness, in contrast to the boundaryless unconscious and the objective psyche. The ego also has a discriminatory, differentiating faculty, equally essential to conscious functioning; the ego differentiates between internal and external, above and below, open and closed. Forces are at work on the ego from all sides. As Jung says, the Self manifests itself through the ego. But other forces also seek manifestation by means of the ego, trying to take their place, or sometimes to overtake the ego, such as the collective consciousness—norms and habits, laws and prejudices. We may particularly see these forces at work in an ego swapped away by a charismatic leader or totalitarian ideologies, and prominently in psychotic masses.

Such forces shape our personality, and if they are too powerful, less psychic room is available in the limited or inattentive ego for the manifestations of the Self. An ego, inundated by media, computers, and cell phones, busy many hours a day with Facebook and Twitter, will have less patience for quiet reading and patient reflection. Utterly meaningless push notifications that ring and light up and demand immediate attention, putting us on constant alert, as if we are in a constant state of emergency, prevent us from being attentive to the Self, the archetype of meaning. As David Brooks points out:

> Technologies are extremely useful for the tasks and pleasures that require shallower forms of consciousness, but they often crowed out and destroy the deeper forms of consciousness people need to thrive. ... Online is a place for human contact but not intimacy ... a place for information but not reflection.[173]

---

[173] David Brooks, Just How Evil Is ech? In *The New York Times Opinion Section,* Friday November 24, 2017, p. 13.

The deleterious effects of technology filter down to the youngest among us, having its effect before a healthy ego has begun to form. Chris Rowan addresses the effects of the widespread use of technology on children's brains:

> Hard-wired for high speed, today's young are entering school struggling with self regulation and attention skills necessary for learning. ... Diagnoses of ADHD, autism, coordination disorder, developmental delays, unintelligible speech, learning difficulties, sensory processing disorder, anxiety, depression, and sleep disorders are associated with technology overuse, and are increasing at an alarming rate.[174]

In the modern era, as we have witnessed in the 20th century, the Self was sometimes projected onto the Authority, at the detriment of the individual's consciousness, which became blurred by being swept away by mass movements and ideologies. In fundamentalism, the ego is in archetypal identification with a perceived transcendent principle, in complete and literal submission to the one and only Divine Principle.

The situation in the postmodern condition is entirely different: The ego-Self axis has been weakened, nearly broken, increasingly replaced by a shadow-persona axis, as will be elaborated below.

Fundamentalism and the postmodern condition seem to be opposites, but they share an uncannily common ground. We need to be aware not only of the shadow of fundamentalism but also of the shadows of the postmodern condition that reside in our souls. Although a fundamentalist worldview is distorted by its very adherence to "absolute truth" and the single-mindedness of totality, postmodern narratives do not rely on truth and validity,

---

[174] Chris Rowan, The Impact of Technology on the Developing Child, in *The Huffington Post,* Dec. 6, 2017: https://www.huffingtonpost.com/cris-rowan/technology-children-negative-impact_b_3343245.html

reality and reliability. The tampered-with, Photoshopped image is at least as attractive as one that has not been interfered with, and the appeal of the image, its attractive power, becomes more important than truth and reality. Neil Postman reminds us that, "One picture, we are told, is worth a thousand words. But a thousand pictures, especially if they are of the same object, may not be worth anything at all."[175]

On the psychological level, healing requires the restoration of the individual meaning, the myth and the story of meaning—not as a reverse to the claims of universal values based on external and/or divine authority, but on the personal level, as elaborated by Jung, Neumann, Edinger and others.[176]

---

[175] Postman, 1993, p.166.
[176] Edward Edinger, *Creation of Consciousness: Jung's Myth for Modern Man*, Toronto, Canada: Inner City Books, 1975, pp.9-12.

CHAPTER

10

Transience and the Transient Personality

Transiency pertains to a sense of being unsettled, of remaining only briefly in one place, or something that lasts only for a short time, something that passes and moves on. Transiency implies easily crossing boundaries and changing shape rather than permanence and constancy. It refers to transient subjects rather than the constancy of objects and is a principle, core feature of our postmodern condition—insofar as we may speak of cores and principles, since *principally*, the postmodern condition is beyond principles.

What are the characteristics of transiency and of the Transient Personality, who seems so well-adjusted to this condition that we cannot really speak of a Transient Personality *Disorder* (TPD), even though, as an enduring pattern pervasive across situations, transiency may certainly pertain to a personality disorder.[177]

---

[177] Helen Marlo, personal communication.

In transiency, we find the following characteristics:

Always Online—Never Offline

It is characteristic of the transient condition that it is never offline. Our phones, tablets and computers buzz, beep, and ring to make sure we are alerted to all incoming messages, emails, calls, postings, and news blasts. Our attention is on alert for these flashes, and it is almost impossible to resist their beckoning. We are wired into the World Wide Web through all our devices that are now extensions of our lives. Conversations are interrupted, smiles from our children are not seen, our concentration remains shallow, and whenever it does deepen, it is quickly brought back to the surface with a ping.

Where I live, it has become common in many shops not to turn the sign "open" around, daringly to declare "closed," closed for the day. This is just another small "sign" of the ego's difficulty to fulfill its task of differentiation, of making that essential distinction, which is its task, to differentiate between open and closed—is that not what we train children to do at an early stage as regards body-ego function? And it reflects the difficulty to let go, close the shop, close the computer, differentiate between work and rest, day and night, sleep and waking hours, everyday and the Shabbat. This anecdotal example may be simple, but I do believe that a blurring and loosening of ego-boundaries, because of reasons I have already pointed out, is taking place, which influences the distinction between I and Thou, thus also the dialogue taking place between people.

Speed without Digestion

We move by car, train, and plane at increasing speed through physical space, but the time it takes to travel cyberspace approximates zero. Seven seconds is the most that the average person will wait for a site to upload before he or she prefers to move on.

The Transient Personality will be quite capable of making quick decisions without necessarily going to great depths. Although the Transient Personality is good at coordination and motor skills, at pushing the buttons and turning the joystick at a speed that makes most people my age dizzy, he or she will often find even simple calculations quite difficult, always in need of a calculator to relieve the brain from mathematical overload. This is only one indication of the shifting emphasis of ego functions: speedier decision-making and greater motor skills, but a weakening in reality testing and perception of self and other as "whole objects." Speed reduces the capacity for *digestion, internalization,* and *reflection.* To stop for a moment, to slow down and listen to the violin of Joshua Bell, at *rush* hour, becomes the abnormal behavior.[178]

Fleeing the Center

By association we flee the center—that is why Freud suggested free association as a technique to escape the bonds of ego consciousness. The freer our associations, the more able we are to break away from the rigidities and the authoritative commands of collective consciousness. But we may lose balance, and perhaps ourselves, as we associatively follow the hints into the endlessness of cyberspace, deconstructing the structures of convention.

At increasing speed, we associatively move centrifugally, i.e., literally "fleeing the center." This leads, in my opinion, to three major consequences:[179]

1. Firstly, we constantly need to locate each other. When calling someone's mobile phone, we rarely ask, "Do I disturb you," but rather, "Where are you?" In transient space, I tend to lose track of where I myself really am.

2. Secondly, contact and relationships tend to spread out on a flat surface of superficial sharing or chattering, "Hey, are you there? I am doing this, what are you doing?" By so-called "friend trackers" you can engage in the stupefying

---

[178] Weingarten 2007, W10.
[179] Cf. Shalit. Jung Journal, p. 93

and time-consuming activity of tracking how other people are engaged in the stupefying and time-consuming activity of tracking you tracking them.

Is it not wonderful that we can chatter and tremble and Twitter and know that Jessica is going for lunch, and someone just missed the bus, and philosophical James throws his pearls of wisdom that "nothing is more frustrating than silent failure."

He is right: Silence may wake up the anxiety, which is put asleep by all the noise. No wonder that studies show that the use of social networks—or, as I see it, anti-social networking—is addictive and affects performance in the real world, for instance lowering student grades.[180] Privacy, on the other hand, is on the decline: Where is discretion when everything is swiftly forwarded? One's life in images, postings, writings are stored forever, out of reach, out of view, but always there. Where is the poet's private song, handwritten and learned by heart? What will be the fate of secrecy when it can no longer be *set apart*, which is the etymological meaning of secret—i.e., identical to the origin of *kadosh*, the word for holy in Hebrew? Often people come to therapy to find *privacy*, and in privacy find an outlet for their personal poetry. In the analytical temenos, we may serve as secretaries, keepers of holy secrets, especially in a world where everything is known and visible.

3. Thirdly, due to the centrifugal movement away from the center, there is neither a place for depression nor a place for deep impressions—as if no hieroglyphs, i.e., the sacred imprints of constancy and discretion, privacy and an integrated narrative, remain.

---

[180] Aryn Karpinski, xxx.

Memory Deleted

What will be the fate of memory when we gradually split off from the past, from whence we come, and when the ease with which we push the Delete button reflects the Weltanschauung of the Zeitgeist? Not only is *delete* as easy as the imitation of "copy and paste," but do we manage to do much without the magic of a wizard?

We only need to push next, next, next, and finish. Don't even make the effort to read what it is that you sign you agree with, and thank God there is no need to put your signature—from Latin, to mark, the matrix of a seal—in handwriting, which anyway is becoming a forgotten skill of the past.

In order to re-member and re-call, to hear the call of vocation, the narrative of individuation, we may need the compensatory centripetal force. We may need to literally "seek the center," not only for introversion of libido, but also for the sense of continuity.

Is it not ironic that in the city of Chandigarh, conceived by Le Corbusier (the flattener of Paris), the city that Nehru proclaimed should be "unfettered by the traditions of the past," the avant-garde furniture designed by Pierre Jeanneret, found its way to the junkyard? Government clerks had already forgotten the value of the postmodern treasures they were sitting on; treasures aimed at breaking away and forgetting the fetters of the past.

Deconstructing the memories of the past shatters the images of the future. To hear the call of the future, we need to recall the voices of our ancestors. The relationship between memory and mass man will be further elaborated in the chapter on recollection and recollectivization.

Remoteness from Reality; the Vitro Life

Remoteness from reality impairs one's awareness of the consequences of one's actions—an essential aspect of the ego's reality testing. When the car in the computer game overturns, there are no consequences, just reset and start—no pain and no suffering, no harm and no wounds, no guilt and no responsibility, no depth of feeling and no feeling of relatedness. There is only the excitement of instinctual satisfaction. The true danger emerges when the individual comes to relate to reality as if it is merely a computer game.

Remoteness from reality also makes us less concerned with the other as a whole object, so the other becomes an easier target for archetypal projections, relieving us of feeling "unnecessarily" burdened by empathy and compassion. Remote from human reality, reality is as easily distorted as it is in the fundamentalist's demonization of the other. For the fundamentalist, there must be no *other*, and in our post-modern world, we are abandoning the embodiment of reality, turning from in vivo, being in life, to vitro, the artificial life, living a life in a glass, or behind the screen.

In either case, it may be all too easy to drop the bomb or to press the button.[181]

Nonlocality

This condition of nonlocality, of nonplaces, as described by French philosopher Marc Augé, is reflected, as well, by the highways that bypass the towns and the villages, with only signs of the existence of historical monuments. The identity of a place has been replaced by global replications, such as McDonalds and IKEA. Italo Calvino describes this condition so well when he tells us about the city of Trude:

> If on arriving at Trude I had not read the city's name written in big letters, I would have thought I was landing at the same airport from which I had taken off. The suburbs they drove me through were no different from the others, with the same little greenish and yellow houses. Following the same signs we swung around the same flower beds in the same squares. The downtown streets displayed goods, packages, signs

---

[181] Cf. Shalit, 2004: 151ff.

that had not changed at all. This was the first time I had come to Trude, but I already knew the hotel where I happened to be lodged; I had already heard and spoken my dialogues with the buyers and sellers of hardware; I had ended other days identically, looking through the same goblets at the same swaying navels.

Why come to Trude? I asked myself. And I already wanted to leave.

"You can resume your flight whenever you like," they said to me, "but you will arrive at another Trude, absolutely the same, detail by detail. The world is covered by a sole Trude which does not begin and does not end. Only the name of the airport changes."[182]

I believe that the postmodern condition of transiency is a reaction to excessive belief and trust in deceptive authority with its horrendous consequences, to which we have had to bear witness, as well as a result of living under the threat of humanity's extinction, which often entails a search for redeeming rituals.

Musical cacophony and the fragmentation of movement in postmodern dance seem to reflect aspects of transiency, but by its mere performance, at a set time and location, dance enables relationship to the experience. This is also the case as regards the plastic arts and theater, even when the boundary between stage and audience, representation and beholder, is blurred, as in some installations. The intentional replication of the postmodern condition turns it into ritual. The choreographer Yasmeen Godder describes her very atonal work as a quest for a process "from the plasticized and artificial to a search towards the 'authentic,' if such exists."

---

[182] Cf Shalit, Erel. *The Cycle of Life*, 2011, p. 38. In Italo Calvino, *Invisible Cities*, London, England: Vintage, 1997, p. 116.

Photographic Reality

We are increasingly living a photographic reality; photographing and being photographed. There are more than four million closed-circuit cameras in the U.K., most of them in London. That is, everything done in public, and increasingly so in private, is being photographed and possibly watched. This is not only the dystopia of a brave new world of persecutory surveillance, no longer the dividing Iron Curtain of long ago, but now the virtual curtain is spread out, just a few meters above our heads, enabling constant observation. It is the creation of a dual reality: We walk in the street, but we are no longer merely walking, we are being recorded as we walk, and we become increasingly aware that we are being visually recorded, i.e., photographed.

This is not the sense of consciousness that comes from self-observation, of being conscious of our acts, consciously relating to what we do, but creates a field of alienation, because even if we had no criminal or questionable intention, we become aware that we are being observed for such behaviors—no, simply, we are being observed. Susan Sontag said it well, by now many years ago, when she said, "Essentially the camera makes everyone a tourist in other people's reality, and eventually in one's own."[183]

To be a tourist in one's own reality, a tourist in one's own life, is being estranged from a sense of Home, from being anchored in oneself, in one's Self. Let me quote another passage from Susan Sontag's important book *On Photography*; she writes, "Photography implies that we know about the world if we accept it as the camera records it. But this is the opposite of understanding, which starts from *not* accepting the world as it looks."[184]

By the way, her close relationship with photographer Annie Leibowitz adds pertinence to her sharp observations.

Sontag's point is crucial: Information becomes pseudo-knowledge and false understanding, when it relies on the

---

[183] Susan Sontag, *On Photography*, p. 57.
[184] Ibid., p. 23.

recording camera, rather than being anchored in the Self, the archetype of Meaning. Her statement, back in 1973!, that "having an experience becomes identical with taking a photograph of it, and participating in a public event comes more and more to be equivalent to looking at it in photographed form,"[185] sounds prophetic. It is, in fact, the very opposite of the process of psychization, which is decisive in the development of human consciousness.

Dual reality will no longer mean that we merely *live* an experience, say, a moment of love, a gentle touch, an erotic scent, or, a moment of horror, a harsh blow, the stench of defeated trenches. No, the experience will be simultaneously recorded. A parallel photographed reality will blur the boundary between experience and its registration; registration of reality will impinge upon the experience of reality. The psychological, reflected upon, remembered, and explored experience will become obsolete and outnumbered, since everything is already deposited in the exteriorized memory of the hard disk.

Mirroring and the mirrored life is replaced by the photo-graphed life. **Memory, dreams and reflections** are replaced by **photos, applications, and blog posts**. The former are necessary for the sense of depth of experience; the latter ones *flatten* experience.

Our postmodern, transient reality has a defining influence on many people who, as seen in its full-blown manifestation, do not live an authentic life, grounded in one's inner core, center, secret, attending to one's inner voice of meaning, wisdom, self, vocation, but rather live a camera-reality, celluloid/digital, life.

---

[185] Ibid., p. 24.

CHAPTER

The Transient Personality

I apologize for this somewhat lengthy description, but combined with other features of the postmodern condition, this reality change is an enormous leap in humankind's perception of itself and of the world. It does pertain to the sense of transiency that characterizes this era and condition, in the center of which stands the **Transient Personality**. This refers to someone constantly on the go, restless and pseudopresent, seductive in its appearance, a chameleonic master of changing colors, yet captive of its own adaptability. Without anchor in its own memory and self-recollection, the Transient Personality eventually gets trapped, crashing into its memorylessness and dismembered fragmentation.[186] In fact, such people are at home in the postmodern world, and when this is the case, when the Transient Personality is at home in the world, the Self, which is characterized by depth, constancy, wholeness, and meaning, is in exile.

---

[186] I mostly use the masculine pronoun he to refer to the Transient Personality but it is only for ease of writing. It actually includes he and she and now perhaps the newer formulation, they.

They wear many masks, or personae. As one man said, "I can wear any hat; I can speak to the Prime Minister and he will be convinced that I am fully informed about any matter, and I can speak to any woman and she will be convinced of my genuine love for her, so much so that I believe it myself, even though I don't care who she is—I just read her quickly, she becomes an object in my hands, and then I need to get rid of her before she starts to have demands on me." The Transient Personality has the chameleon quality of the as-if personality, of which this is in fact a sub-category.

For the Transient Personality, transient masks of as-if identities defend him or her against the hardships and the responsibility it entails to be internally rooted, of being anchored in one's own sense of meaning. Thus, this traveler is never fully present, does not remain in one place, and is not rooted either internally or in external reality. If he or she does appear to be present, it is an inauthentic presence, a "photographic presence," as if acting in a film.

While there may be many constructive uses of the camera, both in private as well as collective documentation, arts and entertainment, it is quite different when a person lives reality by means of the camera. Yanir, for instance, had a great difficulty feeling his own presence across situations, living in a sense of depersonalization as well derealization, both with his wife and children, and with people at work—he choose to be a prison guard, perhaps guarding his own sense of imprisonment. He asked his therapist to video-record the sessions, so that after his death, they could be given to his children. "Didn't Shakespeare say that life is like acting on a stage," was his take on the British author, "and my best appearance is in therapy with you," he told his astonished therapist, who wasn't sure if he was joking with her or serious. I don't think he himself really knew.

I would dare to say that the Transient Personality is characterized by an identification with photographed and photographing reality. He or she is present in a way similar to the camera—not really and meaningfully present, yet present with the accuracy of the camera.

Those so-called "Big Brother" reality shows have very little to do with authentic living—it is the reality one chooses to create knowing that one is part of an artificial reality show, in which one's self is not mirrored, but in which one is mirrored by a constructed, artificial reality. The other person, somewhat awkwardly and perhaps disturbingly called an object in psychoanalysis, no longer remains an independent person but is related to as a self-object. As such, those with whom one interacts serve to mirror one's own game, play and acting.

This is the pseudoreality of the postmodern condition and transiency. There is hardly a trace of the overt terror evoked by Orwellian dystopia, but what remains is senseless plastic.

Only after having spoken and written on this subject of Transient Personality did I discover, not a little surprised, that Ezra Pound preceded me by a century. In a letter to Viola Baxter Jordan in October 1907, Pound distinguishes between the real self and the exterior self, referring to the former as that part of him "which is most real, most removed from the transient personality, (Persona, a mask), most nearly related to the things that [are] more permanent than this smoke wraith the earth."[187]

The Transient Personality often lives in virtual reality—things are not "for real" but like smoke, and can always restart. The as-if aspect is there because reality is virtual; the disconnection from basic consistency, boundaries, warmth, motherly instinct, etc. Clearly, there is underlying trauma but it is securely locked away, forgotten as if in tin containers of old 35mm film.

Often the survivor of trauma does not manage to remove him- or herself from the event, to wash the traumatic experience

---

[187] Quoted in Peter Liebregts, Ezra Pound and Epiphany, in *Moments of Moment: Aspects of the Literary Epiphany*, 1999, by Wim Tigges, p. 235.

down the drain of the psyche's sewage system. Even the pipelines of our dreams are often too narrow in PTSD to flush away the horrors and the fears of the trauma along with the detritus and the residues of the day. The trauma overflows and inundates us with unrest. The trauma bounces back in the repetitive dreams of post trauma, seemingly symbolless; with the endless repetition of the horrors of trauma; *repetition* rather than *representation*.

While depression, anhedonia, anxiety, hypermnesia, guilt, and grief are some characteristics of the survivor syndrome, as originally formulated by William Niederland,[188] the Transient Personality has a chameleon quality of adjusting to temporary masks as well as a greater capacity for change. Often lacking depth and constancy of feeling, such as genuine love and true friendship, awe, pain, and sadness, restlessly moving on, fleeing the angst, the Transient Personality searches for uncommitted freedom, yet, at the same time, seeks ever-elusive closeness, often through internet dating networks and babble-chats. He or she is less capable of considering the *consequences* of his or her acts because a machine-like quality has become an integral aspect of interpersonal relations. Yet, the Transient Personality has an increased need for instant gratification, inducing his or her acts with narcissistic importance, because words are not spoken discretely in dialogue with another person, or with a community of people whose faces one is familiar with, but with the impersonal, unnamed mass of anonymous and pseudonymous, by means of the machinery in cyberspace.

The Transient Personality rarely turns up for therapy because he does not stay in one place. Constantly on the go, he is actually stuck in his own impermanence, avoiding the soulful quest of the true wanderer. If he does arrive at the door of our clinic, we soon discover he is simultaneously shopping for other therapists, and if he remains for a while, he has a reflexive yet unreflective,

---

[188] William G. Niederland, Clinical Observations on the Survivor Syndrome, International Journal of Psychoanalysis, 49: 313-315, pp. 138-140.

compulsive need to answer the mobile phone, which just cannot be turned off, only to say, "Honey, I am in treatment," as if the cellphone serves as a sublimated umbilical cord attached to whatever virtual breast, whose presence at the other end of the electromagnetic field is necessary to confirm one's mere existence. He will sometimes present you with a wealth of flourishing material but leave you with an uncomfortable feeling or sense that he has made it up or copied it from somewhere, and if it is his own, it is an active unconscious or Self trying to surface, but without the ego really attaching to it—it's a persona manifestation of the inner world. The transient personality is never fully present because he is not really touched, and in any case he soon moves on. The nonlocality and temporality of airports, as Temples of Transiency, soothe the restlessness of the Transient Personality and suits him better than the temenos of the therapy room and the analytical relationship, with its hermetically sealed boundaries.[189]

Even in seemingly personal relations, the object, the other, the real person, is not seen. Affect becomes flat, superficial, or faked. One of the reasons for this is the need to avoid pain, and virtuality can retain the illusion of the painless life, while reality, for instance of feeling, imposes pain and suffering, as well as providing an opportunity for true love. The uniqueness of the object is exchanged for the replicability that comes from turning the person into an object. There is no sense of loyalty—neither to people, nor to places or ideas. They travel, concretely and psychologically, with a passport for every purpose.

Everything seems to lie bare in front of the Transient Personality, just as everything is revealed in front of our eyes by unlimited access and disclosure on the screen. Nature, which loves to hide, to quote Heraclitus, is stripped naked. When everything is brought into light, nakedly exposed, we become blind for what lies

---

[189] Cf Shalit. Jung Journal, p. 95.

behind, underneath. Paradoxically, only what lies hidden, can be seen in full.

"Ever since my own body's eyes through my own hand took away the world of appearance," says Oedipus, "I think I have actually begun to see. ... And this world which is incomprehensible to our senses is the only true world which I now know. All else is simply an illusion which deludes us and confuses our contemplation of the divine."

CHAPTER

12

Will Watson's Grandchild Write Poetry?...
And What Tears Will it Cry?

We are all witnessing, as well as taking part in, the enormous leaps in the cyber development of our era. We have handed over the progress (...?) of the world to the brilliant minds of tomorrow, whose bright, nerdy vision is counterbalanced by apocalyptic fears of the future.

The children and the young of today are born into tomorrow's world of multiple realities and a seemingly boundaryless space of associations and virtuality. For example, the temporary constellation of groups, crowds, and masses, the momentary meeting at the city's crossroads, or the incidental minute of waiting for the metro has now been elevated into virtuality. This forces us to reconsider ego-boundaries, reality-testing, soul, and interiority, and to discern the pathologies of our era, such as "photographic reality" and the transient personality.

Watson in the title of this chapter refers to IBM's superior artificial intelligence computer. He (?) can read millions of documents within seconds and is a *Jeopardy* quiz winner, having instant access to more encyclopedias than most of us know exist.

The mysteries of the dream obviously arrive from beyond the limits and limitations of our ego-consciousness, and the stories told us through dreams hold the secrets and the wisdom of the soul, beyond the boundaries of our ego.

Did I say *soul*?

What does that mean?

Well, people of this era as we are, we follow the Zeitgeist, the spirit of the times, and turn to Wikipedia. For a moment we might try to forget those flashing red light bulbs in our archaic mind or guts or wherever they flash, telling us that there is even a Wikipedia entry with a list of all the known hoaxes on Wikipedia, warning us that the dividing line between true and false, expertise and charlatanism has become a very thin, nearly anorexic line these days. So, hastily we "copy and paste," restrained only by that totally outdated psychic faculty called the *superego*, which requires of us to recognize and give credit to the sources we are quoting, that is, Wikipedia in this case. In parenthesis, since Wikipedia is completely free, supposedly totally democratic, with no need to claim expertise, to "copy and paste" from an entry in Wikipedia does not seem to require giving credit, and students today often believe they have done their homework and passed the exam if they "copy and paste" entire entries.

There are, of course, the more ambitious students, who want to prove they have done their serious work, ending up *rogeting*. It comes from the thesaurus and is defined as the practice of students replacing words and phrases in essays they have copied from the internet with supposedly synonymous alternatives in order to disguise their plagiarism.

It just so happens that it sometimes ends up rather awkward. Chris Sadler, a British business lecturer who coined the term, gives the example of finding "sinister buttocks" in one of his student papers. The origin of "sinister buttocks" was "left behind." "[There's] no attempt to understand either the source or target text," says Sadler, so right on the target—and this is part of the problem of our era.[190] There is a lack of understanding that there is need for *meaning*.

Much abbreviated, Wikipedia says about soul:

> The soul, in many religious, philosophical and mythological traditions, is the incorporeal and, in many conceptions, immortal essence of a living thing.[191]

Reviewing the idea of soul, the article mentions the religions, even neuroscience and parapsychology, Kant, Aquinas, and Theosophy. It even says that *Anima Mundi* is "the concept of a 'world soul' connecting all living organisms on the planet." Otherwise, no Freud, no Jung, no psychology. It's like talking about the psyche without the psyche. We should not be overly surprised, since many branches of psychology exclude the psyche from their discourse. The article ends, however, with a brief discourse on the weight of the soul. "In 1901," we read:

> Duncan MacDougall made weight measurements of patients as they died. He claimed that there was weight loss of varying amounts at the time of death. The physicist Robert L. Park has written [that] MacDougall's experiments "are not regarded today as having any scientific merit" and the psychologist Bruce Hood wrote that "because the weight loss was not reliable or replicable, his findings were unscientific.

Even if totally unscientific, the weight of soul is, of course, a tremendously interesting subject.

---

[190] http://www.theguardian.com/education/shortcuts/2014/aug/08/rogeting-sinister-buttocks-students-essays-plagiarising-thesaurus.
[191] https://en.wikipedia.org/wiki/Soul.

In the film *Smoke*, based on a script by Paul Auster, the protagonist rushes into the Brooklyn neighborhood tobacco shop and asks his fellow customers, "How do you weigh smoke?" They don't manage to resolve the riddle. The story tells us that neither did Queen Elizabeth know how to solve it, when 16th-century poet and explorer Sir Walter Raleigh, who introduced tobacco to England, asked her, "How do you weigh smoke?" Clever as she was, she supposedly answered him, "How can you weigh smoke? It's like weighing air or someone's soul."

Wasn't Queen Elizabeth right that we cannot weigh the soul? How can we give the soul material expression? Is the soul not as elusive as the wind? As fragile as a soap bubble? As transparent as glass? Yet, when present, is the soul not as full of wonder as Iris the rainbow, daughter of Thaumas the Wondrous; do we not hear the Voice of the soul on top of every mountain, its echo deep in every valley, its power in every wave that hits the shore? It is the soul that gives character to the wrinkles of old age; when the spirit is lost, the aging wrinkles turn into parched furrows.

Raleigh's answer to the question of how you weigh smoke was to weigh the cigarette, smoke it, weigh the ash, then subtract the weight of the ash from that of the cigarette, and there you have the weight of smoke. When we deduct the weight of the remaining ashes from the weight of the unsmoked cigarette, we realize that soul and spirit weigh more than matter, but also that they need the matter of the cigarette paper as a container—otherwise they simply disperse into thin abstractions.

That is, I believe, essential. We need nature, instincts, and the embodiment of actual reality, in order to give a dwelling to the soul, a dwelling without which the soul dissipates. In Jewish legend and belief, God needs the Shekhinah to be present. Shekhinah means *dwelling*. So however abstract God and even the *image* of God may be—as for instance the God of Spinoza "who reveals himself in the orderly harmony of what exists, not in a God who concerns himself with the fates and actions of human beings,"—the transparency of soul requires the solid container of what we might call reality.

And if the soul expresses itself by our individual images, is it then perhaps our imagination that liberates us from the bonds of predetermined fate? The images of our interiority radiate from the soul. Evil resides where there is a lack of imagination, said Hannah Arendt. And Jung says, "*We* have no imagination for evil, but evil *has us in its grip.*"[192] That constitutes a condition of soullessness. No less is an absence of imagination, symbolization, and a loss of soul the post-traumatic *consequence* of evil, as we know from many a Holocaust survivor. The existence of the soul, that ungraspable, purely elusive idea of anima, whether in man or woman, cannot be confined by earthly empires, neither by rules nor by imperatives, but can only be poetically imagined, for instance, as the soft light of the moon or that image of a mirror that mirrors the image. As James Hillman so clearly says:

> It is as if consciousness rests upon a self-sustaining and imagining substrate—an inner place or deeper person or ongoing presence—that is simply there even when all our subjectivity, ego and consciousness go into eclipse. Soul appears as a factor independent of the events in which we are immersed. ... That unknown component which makes meaning possible, turns events into experiences, is communicated in love and has a religious concern. [193]

Just as dreams are the myths that we are told individually by the storyteller in our soul, so society, civilization, science, and technology, the development of humanity, and the era in which we live are themselves manifestations of our collective, oneiric myths, that is, they are manifestations of our collective dreams and images. Life is the story of the unfolding of the Self, and as Jung has repeatedly reminded us, the developments and the changes in society are manifestations of the unfolding of the objective psyche.

---

[192] *MDR*, p. 331.
[193] James Hillman, *Re-visioning Psychology*, Harper Perennial, 1992, p.xvi.

So, what is this world we have conjured up, based on the dreams, desires, and defects of our minds? And what will the world look like, the world in which our grandchildren or great-grand-children will grow up to become adults? What will society look like, what will be the character of civilization, what will be the nature of nature, the relationship and balance between Eros and Logos? What might be the essence of humanity, of the human individual, and human relations in the world that is taking shape, the contours of which we can perhaps begin to discern, as the future approaches us through the mist of the unknown, on the hazy bridge of anticipation and apprehension?

I'm sure many of you have likely contemplated these issues. As clinicians, we encounter individuals, sometimes a family, or a group. In clinical praxis, confronted with individual pain and suffering, we rarely deal directly with the greater picture of the times and the era in which we live, though it may dwell in the background of our mind.

However, consider the fact that now the estimated lifetime prevalence of PTSD, of post-trauma in the U.S. population, is about 10 percent and about 10 percent of children ages 4-17 are diagnosed with ADHD. If, as Freud taught us, the abnormal tells us something about the normal, then *trauma* and the difficulty to *focus* and *stay in one place* are phenomena that tell us something about our era.

Trauma is axiomatically a *result* of environmental impact—this is what Freud debated when he oscillated between his trauma theory versus fantasy and imagination. And ADHD, whatever its etiology, *impacts* the environment.

Many parts of our world suffer from collective post-trauma. You only need to look at several Arab countries today, in which entire populations are being traumatized.

And we suffer post-trauma from Auschwitz and Hiroshima, and their impact on our collective psyche. They have also made us restless, fearful from remaining centered within ourselves, not only because of *Attention Deficiency*, but because of an *Abundance of Tension*. The present waves of migration are horrific, external manifestations of trying to escape loci of unbearable tension. I believe that just as we are affected by pollution and ecological occurrences, we internalize, to varying individual degrees, collective human phenomena.

Jung says, "Neurosis is intimately bound up with the problem of our time and really represents an unsuccessful attempt on the part of the individual to solve the general problem in his own person."[194] That is quite a radical statement. He claims that personal suffering is an attempt to solve the problems of our time!

Jung was troubled by the transition in which man gets inflated, seeing himself as "lord of the earth, the air, and the water, and that on his decision hangs the historical fate of nations. ... In this reality man is the slave and victim of the machines that have conquered space and time for him; he is intimidated and endangered by the might of the military technology which is supposed to safeguard his physical existence..."[195]
And, says Jung:

> The degenerative symptoms of urban civilization have already led to a widespread though not generally conscious *fear*, which *loves noise* because it stops the fear from being heard. Noise is welcome because it stops the inner instinctive warning. ... Noise protects us from painful reflection, it scatters our anxious dreams.[196]

Noise is a central feature in our exposure to an abundance of distractions. Researchers claim that when working on a computer screen, people are distracted at an average of 37 times per hour.

"As scientific understanding has grown, so our world has become dehumanized," says Jung, and continues:

---

[194] CW 7, par. 18.
[195] CW 10, par. 524.
[196] Adler,1975. *C.G. Jung Letters: 1951-1961*. Princeton, N.J.: Princeton University Press.

> Man feels himself isolated in the cosmos, because he is no longer involved in nature. ... Natural phenomena ... have slowly lost their symbolic implications. Thunder is no longer the voice of an angry god, nor is lightning his avenging missile. No river contains a spirit, no tree is the life principle of a man, no snake the embodiment of wisdom, no mountain cave the home of a great demon. No voices now speak to man from stones, plants, and animals, nor does he speak to them believing they can hear. His contact with nature is gone, and with it has gone the profound emotional energy that this symbolic connection supplied.[197]

Jung's very description, his imagery and symbolism, is full of soul and spirit, whether we know how to define that or just accept that we sense it. Not only have "natural phenomena lost their symbolic implications," but nature, inner and outer, is necessary to generate the symbolic reality of the psyche.

I have chosen to use the idea of *Watson's grandchild* as the image that I can hardly imagine to characterize a future that most of us can only speculate about, a future which quite a lot of brilliant young men and women, and not a few whom we might consider quite disturbed, are engaged in shaping.

If you google *Watson*, it will take pages until you get to either Thomas Watson and the telephone, or to John Watson, the founder of Behaviorism. You mainly get to the IBM Artificial Intelligence supercomputer, which goes by the name Watson, which, according to IBM, is "a cognitive system enabling a new partnership between people and computers." *Partnership* sounds nice and friendly, and I am sure Watson does and can do a lot of good, but what kind of partnership?

Watson, supposedly a "he," has sadly won against the champions of the *Jeopardy* show. I don't know if those human

---

[197] Jung 1964, *Man and His Symbols*, p. 95.

champions, when admitting defeat, saw it fit to quote Sherlock Holmes from Arthur Conan Doyle's book *The Final Problem*, when Sherlock says, "You know my powers, my dear Watson, and yet I was forced to confess that I had at last met an antagonist who was my intellectual equal." The defeat to *this* Watson may, however, just have been too jeopardizing.

To me, I must admit, Watson's triumph is sad. Not because the machine does better than the human—that is, of course, the very reason why we build machines: The car moves faster than an old man walks, and the calculator counts both faster and with greater precision than the young person computes. It is because many of those who shape tomorrow's world fantasize about the machine as if it were human, and when doing so, humans tend to become all the more machinelike. Or, rather, the postmodern machine, the computer, is induced with capabilities far beyond computing and the processing of data. We project human traits, the posthuman self, and divinity into the data processor. Thus, some become, or believe themselves to be the creators of the divine master, while most of us become its benefactors as well, but also its victims—for instance, reduced to a collection of data in a global archive managed by a few global megacompanies in virtual reality, such as Google, Facebook, Amazon, and IBM's Watson.

When, for example, do I need an app to *tell me* that I am happy, as one of those computer app designers happily declared he is working on? Does that not paradoxically mean that I *am* not happy, since feeling axiomatically needs to be felt? Can, in this postmodern world, feeling exist without being felt from within? Can I *be* happy by being told so by a nonhuman app? What happens if we do accept that this is so—that I consider myself to be happy if I receive the information, based on a collection of assembled data? Where is the locus of control? This, by the way, is what happens with the GPS, with navigating systems, which serve many so well—it becomes addictive, and many come to rely on it even when they normally would know the way. By using it, it increases insecurity in one's perception of the external map. With the "happiness app," insecurity is created as regards *self-perception*.

The black box of behaviorism turns into an increasingly empty box that is fed with increasingly external data, decreasingly fed back what is uploaded from within but fed forward by downloaded apps, some of which are more traps than apps. Thus, in this brave new world some, central aspects that we consider human become obsolete. The thought that humanity will become obsolete is somewhat repulsive because I don't believe that Watson's grandchild will have a soul.

I asked a senior manager at the IBM project if Watson's grandchild will be able to write poetry—and what is more soulful than dreams, tears, and poetry! Will it have the capacity of moral judgment other than what it has been fed—and it has been fed by those very people who without the depth of questioning feed it with a grand Self-projection, with the idea that a data-collecting device is the Master of the Future—and they are its creator.

He claimed it won't write poetry, other than mechanically. *Mechanically* is, of course, the very opposite of *poetry*. So, it won't be induced with soul. If he is right, that may be a certain relief. But my concern is "the new partnership" between human and machine, which considerably alters the human.

In September 2015, at the online news site *Monitor Daily*, under the headline "Watson, you're improving," we could read that "ever since he was designed, Watson was destined for greatness" and that, among other things:

> Soon Watson might be able to find out what you were cooking last night and what Christmas presents you gave your girlfriend 5 years ago. Watson is growing stronger with every bit of information he gets and we all know that strength is power. ... Now it looks like Watson is going to be able to see everything about everyone from everywhere.[198]

---

[198] *Monitor Daily*, September 2015, Watson, you're improving. https://www. themonitordaily.com/watson-youre-improving/26624/

It is concerning that Watson is "going to be able to see everything about everyone from everywhere." What does that imply two generations ahead? It does create a "brave new world." In his foreword to *Brave New World Revisited*, Aldous Huxley writes, "Life is short and information endless: nobody has time for everything."[199] If we are tied up to the idea of Watson, that is no longer true, and when that no longer remains true, there is no need to choose and no need for decision making. That is, some of those features we consider essential to the ego become obsolete. We can already witness how finger coordination as we tap the phone or tablet is more essential than differentiating the external from the internal, an otherwise recognized crucial feature of the ego.

Additionally, the accumulation of all-inclusive data, with all its benefits, creates a double reality. It creates a dual world—the unmediated direct experience of reality and its double, the data-collected, recorded reality. They are, in fact, already merging, as I already mentioned, in the photographic reality of the Transient Personality, which signifies the harm caused to the unmediated experience of human reality.

Modernity versus Postmodernity

The modern era signified a break from the past and the opening of the gates to individuality and progress. Yet, as Pete Seeger and the Byrds sing and the Bible says, to everything there is a ... shadow—actually a season that turns into shadow as life turns, turns, and keeps turning. Or in the words of Theodor Adorno, "Enlightenment, understood in the widest sense as the advance of thought, has always aimed at liberating human beings from fear and installing them as masters. Yet the wholly enlightened earth radiates under the sign of disaster triumphant."[200] In modernity,

---

[199] Aldous Huxley, *Brave New World*, in the foreword.
[200] Theodor W. Adorno, *Negative Dialectics*, translated by E.B. Ashton. Seabury Press; New York, NY: London, England: Routledge, 1973, p. 210.

Man—capital M and generally speaking, male—is the Master. In post-modernity, Machine is the Master.

The great time of enlightenment meant repression of the unconscious and of nature. It was Freud who rediscovered repressed nature, the sex behind the layers, as it were, of collective consciousness, of a repressive superego, nature and instinct that had been sacrificed at the altar of collective progress. With the repression of nature, inner and outer, personal and collective, comes alienation and loneliness. Thus, early-20th-century literature, prominently Kafka, pertains to the sense of alienation as a consequence of the dehumanization, which was a result of humanity's attempts to humanize the world. With its shadow of alienation, loneliness, and estrangement, the modern condition relied on people—as a workforce and as consumers to keep the machines of modern times going, as soldiers to die in the trenches, and civilians to be exterminated. People were needed for the mass political movements. In order to carry out their programs, Stalin and Hitler needed people who were caught up in archetypal identification with the leader, with the cause and the divine goal of Nazism, Fascism and Communism—in other words, masses that marched.

In contrast to the alienation and loneliness, the Movement, with a capital M, provided a sense of belonging, a greatness greater than oneself, a grandeur beyond one's littleness. Carried away on the wings of narcissistic inflation. It is the average man, no less than the cruel beast, that becomes the willing executioner of the dehumanized enemy.

Even today in our postmodern world, many of us wonder why ISIS attracts young people from the West in its barbarian activity of ruthless war and strife. The pattern repeats itself because, as Heraclitus said, "War is the *Father* of all things." War is certainly not the *Mother* of all things, but war, strife, martial energy, are no less powerful than Eros. Based as it is on establishing borders, boundaries, and differentiation, when identified with, war becomes the means of *destroying* the Other. Because in identification with any elevated archetypal idea or phenomenon, the bad, the rotten, the evil shadow must be carried by the enemy-other. The illusion is then easily created that by destroying the evil other: I shall transcend the mundane, and peace on earth shall reign in the Reich of a Thousand Years; or there shall be universal brotherhood of man and always for utopian eternity. Shadowy dystopia rather than blissful utopia seems to be a healthier alternative.

Jurgen Todenhofer, a German journalist who spent time with ISIS, claims that the militants planned to kill hundreds of millions of people by nuclear annihilation, preparing, as he says, "the largest religious cleansing in history."[201] So while some swing high up in the sky to the energizing fantasy of reaching the Gate of God, the rest of the world may live in denial and dissociation, or fall into despair and depression.

Nearly 40 million people were killed or wounded in World War I. The Battle of the Somme alone was devastating, resulting from the introduction of tanks:

> One of the bloodiest battles of the First World War, almost 20,000 British troops died on the first day alone, and by the time the fighting was over, 419,000 British and 204,000 French soldiers had been killed or wounded; German causalities totaled some 465,000. Vice Sergeant Hugo Frick wrote to his mother: "This is not war, but a mutual annihilation using technological strength.[202]

More than 60 million were *killed* in World War II. To me, so far, Auschwitz stands as the monument of the shadow of cold evil of the modern era. And how easily more than those astounding numbers could be killed by the atom bomb, which marked the transition into the postmodern era. Now we all know that the most

---

[201] http://www.washingtontimes.com/news/2015/sep/28/l-todd-wood-isis-planning-nuclear-tsunami/.
[202] Volker Ullrich, *Hitler Ascent, 1889-1939*, Kindle Version, loc. 1570.

civilized of nations can carry out genocide by perfection, and humanity can destroy itself entirely, quickly, and from afar.

Just like Major Claude Eatherly, one of the Hiroshima pilots, who suffered the torments of his conscience and attempted suicide. He was hospitalized and diagnosed as having lost his connection to reality, but who can blame him? Who can be a witness to the unbearable and bear it? Minimizing such an overwhelming experience with a simplistic phenomenological diagnosis is not the answer. In fact, it reminds me of how in the last few years several people from villages in Israel close to the border with Gaza were treated with anti-hallucinatory drugs after complaining that they heard digging under their houses. However, sometimes we need to hear the very real sounds from below—the tunnels that had been dug in preparation for attacking those villages would have been discovered much earlier, perhaps by listening to the stories people told rather than silencing them with pills.

Humanity continues to play with fire, to use it for both creative and destructive development. The transformation of the elements by heating is an archetypal act, which with the help of Prometheus as an image, was handed over to the sphere of the human ego, so that we now very easily can turn on the stove for cooking or smelting. We cannot, and probably should not be constantly aware that there is an element of the gods, of the archetypal world, present in these everyday acts. Danger comes, however, when we lose proportion, when we no longer contemplate the balance between the forces at work. In the nuclear explosion, we cannot fail to see and be aware of the apocalyptic potential of humanity's self-destruction.

Our postmodern world is pushing us to an increasing distance from reality, which means that reality does not necessarily feel real. It is important to be aware of the boundary between fantasy and reality, which is a central function of the ego. This boundary is increasingly blurred by the constant inundation of fake news, retouched images, virtual reality.

It is now a world of global inter-connectedness. With the so-called Internet of Things—the connectivity of devices, systems, and services, by which anything from biochip implants, smartphones, cars and planes and entire cities—in another few years, in 2020, 50 billion things, at least, will be communicating back and forth on a global scale, receiving and delivering information. We have recently seen how software in cars can be manipulated, and with global interconnectivity, most anything can be activated—or shut down—from afar.

Noise is another factor we live with. There are those who withdraw from the overexposure to noise and city life. An intelligent young man whom I know could not bear to work in his father's business of supplying equipment for weddings, which I believe in Israel often is a particularly noisy activity. He withdrew to a farm in the desert, spent his time taking individuals on camel rides for stretches of time, tending to the animals, with whom he communicated more freely than with most people. This is the Borderland Personality that Jerome Bernstein so eloquently writes about.[203] At the other end of the spectrum, we have the Transient Personality, whom I have already described, and is in many ways the opposite of the person suffering from survivor syndrome and certainly opposite from the Borderland personality.

Jung said that many of his patients did not fit into any particular diagnosis, but their neurosis pertained to a sense of meaninglessness. When the Transient Personality experiences a sense of meaninglessness, that is actually a sign of waking up, and healing might be possible. The genuine experience, of sadness, pain, loss, meaninglessness, is, to quite an extent, the healing itself.

I believe that in order to hold the balance better, in a derealized world in which we have projected the Self not only onto but into the screen and the machine and sometimes weapons, we need to listen to the stories that the gods tell us through the Self in our soul, to the big dreams that serve not only the individual, but speak to the community we live in as well. They offer up images of the state of our world and the distress we experience, offering us more than a pill. It is up to us to pay attention. The following dream

---

[203] Jerome Bernstein, *Living in The Borderland: The Evolution of Consciousness and the Challenge of Healing Trauma*, London, England: Routledge, 2005.

was told by a young woman, a social worker in a community dream group. She has given me permission to share it:

I'm in a house or an apartment which is kind of open. All my friends are there.

Almost no walls all around, everything is completely open. But still I know it is my place, my house.

The police come. They are searching for something. They are very determined, very aggressive, and want to see what I am hiding.

I know I'm not hiding anything, I know that I'm completely clean, but still they are searching for something. They search all possible wardrobes, and there are shelves there that they particularly search.

The shelves are empty, there is nothing there.

Then a second scene, really surrealistic.

It is really strong, almost a physical experience. I feel as if there is **a rain** coming to my forehead from above. This rain is extremely strong. I almost feel as if there are arrows coming into my head.

(My forehead is ATTACKED with a huge amount of **arrows**. It is a toxic rain with sick red-orange transparent greenish color! And it's hitting my forehead directly.)

And I know that this rain is toxic, and I know those arrows are very toxic. And I know that this is really bad for the environment. And I know that I have to run away, like to run for my life or something. And everyone else has to run (it's Apocalyptic).

The whole ground is affected by this rain. It's almost like you shouldn't be on the ground, you have to run, but **not with your feet on the ground**, somehow. And then I run for a few hundred meters and I finally come to an area or a field, which is more natural or more healthy. And then **the rain** that comes there is also more healthy and more pure, and it's much nicer, light or clear.

So, I'm running still, and I know that I still have to run, and then I come to the area of an industrial zone. There are pipelines, and it reminds me a bit of one of my assignments where I worked near an oil company. I used to take a walk there where all those pipelines were all around. So, now in this area it's really, really toxic on the ground. So, I really have to hold myself on these pipelines, and that's the only way to go. I'm holding my hands on it, and that's the way I go farther.

And my mother is on my left side. And I suddenly become extremely sad because I know or I think that she doesn't have the strength, she is so old, and she cannot manage life anymore, somehow. I am extremely sad, but then I'm thinking about myself. I cannot go on. I feel I'm not motivated. I feel like I don't have enough strength. I'm completely finished.

I realize my mother actually, she has the strength, and she gives me the strength, and she is, kind of, cheering me up, and saying: "Come on, you can do it! You have to go on." And then there was one thought also in between that—I am thinking that I should inspire other people but if I don't have my own energy to go on, how should I do that, that's kind of ridiculous. And then my mother is cheering me up and giving me the strength, and then I just feel so grateful that she is there. ... And then I continue, and I feel: "Ok I can do this, I have the strength, I can go on," and then I realize that she is suddenly losing her energy, so I have to take care of **her**.

So, I feel extreme responsibility suddenly!

And I'm trying to take her over to the right side. But it is all happening on this pipelines, so basically, I should not drop her ... and her body suddenly becomes A HUGE BODY, like it doesn't look like human, it doesn't look like a human body or a woman's body. It looks more like **a goddess or a statue** or something extremely beautiful, and it's big, and I realize I cannot carry this

myself, and then there is a person in the front there, somehow, in front of these pipelines.

It's a **guy**, he is maybe 20-25, and he actually reminds me about one person at work that came at some point to fix some computer problem or something.

And this person is helping me out to carry this body and help my mother to go on.

The dream begins in the private space, which is completely transparent, and intruded upon by the police, by the authorities, which in this case requires the person to be convinced of her clean conscience. The extent to which we are observed and data is collected about us, and we are being surveilled, the extent to which we are losing our privacy, it is difficult to keep a hold of our own inner kernel—call it Self—since it needs to be protected, to be set apart—otherwise its nothingness seems to be nothing, as if there is nothing at all, nothing of value.

The second part or parts of the dream are out in the open. That public sphere has become toxic, with an apocalyptic sense about it. There is a need to flee. It is as if the ground, reality, has become so polluted that it cannot be walked upon—I believe this is an aspect of the collective reaction to the traumata that human-kind has had to suffer so that we now dissociate, disconnect from physical reality. Will the mother, whether personal or archetypal, be able to provide ground, basis, backing, growth, encouragement, reality, or is she too tired by now?

She can seemingly hold out a bit longer, but then, at some point, Mother loses her strength and needs to be cared for. This is where the young computer guy appears, helping the dreamer to carry the Great Mother, helping the Great Mother to carry on her journey.

We obviously need to help the Self, the gods, the objective psyche that speaks through our dreams, to tell their story, so that those who build the future will help carry the Great Mother, earth and nature, the container and the embodiment of reality, rather than merely building virtual Towers of Babel, because the confusion might just be too great for us humans to resolve.

CHAPTER 13

The Battle of Character

No Body versus Sum-1

With increasing ease, we wear *transient masks*, no longer burdening the personae with uncalled-for meaning. That is, we avoid burdening ourselves with the call of the voice of the Self, the archetype of meaning, which may speak through the mask, *per sonare*, by means of voice. In the transient space of the cyberworld, we have passwords, Facebook profiles, blog pseudonyms, usernames, chat rooms, and second lives—an unending charade of masks to hide our true identity, until we reach the point of dissociating completely from being ourselves, truly.

As already mentioned, following the Eichmann trial, Hannah Arendt brought "The Rule of Nobody" to the foreground. I believe that *Someone* and *Nobody* are the combatants on the battlefield of postmodernity. *Nobody* means a lack of body, of substance, no-1, "not-1," whereas Someone adds up to One Whole Sum—Sum-1—*Someone*, i.e., being of one substance; in Hebrew the word for someone, מישהו, means "whom he/she is."

The Nobody speaks Officialese, the pedantic and verbose language often characteristic of official documents. Hannah Arendt reported Eichmann saying, "Officialese is my only language,"[204] the language of the flat words of collective consciousness, behind which the depth of truth and meaning remain hidden—the technical exchange of the width of railway tracks between different countries, without mentioning that the issue is to resolve transportation of humans to the camps of dehumanization.

This is a battle of *character*—of being and having character, which comes from the Greek, meaning "engraved mark," of having character and being a character in one's own drama. Heraclitus promoted the *daemon*, which originally pertained to the share the gods have given to man, into man's psyche, and said, "Man's character is his daemon."

We may understand Pirandello's brilliant play *Six Characters in Search of an Author* from this perspective: "A character, sir can always ask a man who he is," says the father character in the play, and continues: "Because a character truly has a life of his own, marked by his own characteristics, because of which he is always 'someone.' On the other hand, a man ... a man in general, can be 'nobody'."[205] This is our choice, to be *someone* or *nobody*—an author, character and actor, or an unauthored actor without character.

For a *someone*, somebody of and with substance, with character, i.e., with the consistency of engraved marks—by the way, compare *character* with *hieroglyph*, sacred imprint—for a someone to constellate, there are certain prerequisites that have to be fulfilled as regards the ego and the Self: The ego, as center of consciousness and conscious identity, must carry out its central task of *boundaries and differentiation*—what is within and what without, what above and what below, behind or in front; and the boundary-bound ego must be *rooted* in the whole person, in the soul or the Self or the Daemon or character or some center and wholeness beyond the limitations of consciousness.

---

[204] Arendt, p. 48.
[205] Luigi Pirandello, *Six Characters in Search of an Author*, 1998, p. 61.

If these conditions are met, if the ego is simultaneously separate and differentiated, *and* recognizes its limitations, *and* is rooted in and related to an inner as well as outer *Other*, then there is a chance the ego can function as a *third*—such a vital concept in psychoanalysis—as a regulator between id and superego, between shadow and persona. This reflects an ego, which has a vital, dynamic, and living relationship with the Self, the archetype of meaning.

On the other hand, in the absence of the above, in a condition of fragmentation, of speed without digestion, of splitting and disconnection (dissociation, derealization, depersonalization) due to fragmented association (e.g., restlessly moving from site to site on the internet, or from relationship to relationship without forming a real relationship, from Match.com and JDate to Cupid and Tinder), of fleeing the center, i.e., away from the center, away from the Self, away from oneself—then the ego will not be able to fulfill its tasks, neither of differentiation nor as captain of the ship, and not as manifestation of the Self.

Jung borrowed the term, *persona*, for the dress we wear, the exterior garment of our personality, from the mask worn by the actors in ancient Greek theater. But persona, like person, which originally meant a *character* in a drama, comes from the wooden mask in which the mouth would strengthen the sound of the voice, the voice from within, the inner voice.

Prior to Jung, Ezra Pound applied persona to denote a "literary character representing the voice of the author." That is, the mask we wear may have character, the engraved marks, to express the voice, the sound, and the meaning, of the author, i.e., of whom we are, the *someone* in me.

Persona comes from *per sonare*, i.e., by means of voice—the mask of the persona could conceal the everyday face of the actor's ego, to reveal and enable the manifestation of the gods, or what we may call archetypal energies and images, and of our daimon, of one's character.

In the masks of rites and rituals, the ordinary face which everyday serves not only as an external mask of appearance and adjustment, but also as a barrier against the overwhelming realm of the archetypes, is removed, so that trance can be induced and the transcendent archetypes, the gods and the daimons and character can manifest.

But this is not the case in the postmodern world of transiency. Here, the ego-Self axis has ruptured due to the factors I previously mentioned: speed without digestion, fleeing the center, remoteness from reality, forgetting the connecting link. Therefore, one's inner core, or daimon, or sense of meaning, cannot stream freely into the fibers of the personality.

The unrooted ego—more able to coordinate than to reflect—does not serve as a third between shadow and persona, whereby they fuse.

The persona becomes like the actor without character, as in Pirandello's play. The mask, when not permeated by meaning, becomes an empty shell, easily pervaded by shadows and by a charade of pretensions. No longer the wooden mask, it becomes synthetic rather than genuine, a mask of plastic—*plassein*, i.e., "fit for molding," to be cast into any shape, without character.

Nearly half a century ago, Andy Warhol pointed at the self-replicating culture of plastic, perhaps blurring the line between being a critical observer and a willing participant, his art embodying both an eye on the culture of production and the self-reproduction of culture. As Andy Warhol testified about himself, "Everybody's plastic. ... I want to be plastic."

Plastic has its definite advantages, and we can no longer live without it. However, plastic reflects something being synthetic and artificial rather than natural and genuine. Plastic can, as well, be recast out of proportion, but what grows out of proportion is carcinogenic. With the benefits of plastic, and the idea and the cultural attitude that we may call "plastic," come its shadows, such as widespread environmental harm, inauthenticity, imitation, and reproduction rather than individual touch and feeling. Of course, the origin of plastic is our ancestors, dead life transformed into black gold.

An uncanny new development using plastic is in the manufacture of lifelike sex dolls or sensual robots considered to be manifestations of artificial humanity, as they are euphemistically called. They may be the perfect solution for loneliness in our world of transiency, where it is too difficult to relate to another human being. These dolls are big on obedience, no talking back, and only the right amount of aggression. Not alive, not dead, they exist in the gray zone of no existence, never having been born, never able to die, nothing but the mask of persona. Noa Manheim expresses her concern:

> Falling in love with the uncanny, with what is neither alive nor dead, will always be barren, one-sided and incomplete. And when the object of desire is revealed in the fullness of its artificiality, the loneliness becomes more bitter and horrible than ever. The price for yielding to anxiety—of castration, of intimacy, of a bond, of the gaze or the blindness of the other—and turning one's back on what is human, all too human, is loss of self.[206]

In a culture of plastic, the way we appear to the world, our persona, may be infinitely recast, find endless manifestations. In cyberspace, we easily hide behind pseudonyms and borrowed identities, whereby individual morality and responsibility are weakened. There is no function of the *third*, serving as intermediary, boundary, and control. There is no third term, or transcendent function, as Jung called it, to mediate the opposites, to bridge between the real and the imaginary.

For example, a man with a record of harassing women on internet chats dreams that he is locked up in a prison cell. He hears the voice of an old man, who asks him questions that he must answer in order to be released. However, rather than listening and reflecting,

---

[206] Noa Manheim, An Uncanny Kind of Love, In *Haaretz Weekend*, Friday May 4, 2018, p. 8

he tries to escape, merely annoyed at the voice calling upon him. He flees his cell, and as he approaches the prison walls, a woman, whom he identifies as a female judge, tells him he can exit, but only through the merely half-meter-high opening in the wall.

If he truly would listen to the voice calling from within, that requires him to respond (responsibility; *respondere*, to pledge in return), he might wake up to understanding his guilt, the need for boundaries, and attending to his conscience. If he would reflect on the words of the clever and compassionate judge, he would understand that the time has come to bend low, that freedom dwells not in characterless sociopathy but in humble and reflective conscience. Never having heard of Andy Warhol, he says, "We are the plastic generation." The dream images portray his avoidance of meaningful responsibility and guided reflection.

The movie *The Truman Show* portrayed the condition in which everybody participates in the scam of a reality show, where everything is a constructed, artificial, in vitro reality, except for the one, lone *true man*. When criminals become celebrity interviewees, the shadow has invaded the persona. When internet pseudonyms let perversion, abuse, and harassment become everyday commodities, the shadow has pervaded the persona.

At its extreme, a person may sell his or her life, as has been done on eBay, and walk away from it all.[207] Then there is no need to stay in place, to remain and truly be somebody, to be a some-one. By fleeing the center, we run away from being "of one substance," that is, being someone, and what being someone entails: responsibility, integrity, and tragedy.

The result is a weakening of meaningful (i.e., Self-based) identity, toward false and faked identities, carrying the transient traits of as-ifs rather than *being* someone. Consequently, the ego ceases to mediate between shadow and persona, whereby they fuse. In dissociation, aspects of the shadow may be split off, with the ego possibly adapting to the requirements of social appearance,

---

[207] Lewis Carter, 2008.

i.e., the persona, while in transiency, the shadow flows into the arteries of the persona, leading to sociopathy and perversion. One example is the case of Lori Drew, a 49-year-old woman who on June 16, 2008, pleaded not guilty to charges of having created a fictitious identity on MySpace called Josh Evans, by means of which she cyber-befriended her next-door neighbor, 13-year-old Megan Meier. The cyberlove that led to cyberbullying allegedly provoked Megan's suicide.[208]

Who is guilty? Is anyone guilty? Lori Drew was found guilty of misdemeanor and unauthorized computer access. What kind of love and feelings are we talking about when the object is fictitious? It does resemble the teenager's projected feelings and phantasies onto an idol, but the possibility of interaction blurs the boundaries between fictitiousness and reality—in the above case, the unstable young girl did not differentiate between vivo and vitro, reality and cyberspace, or between true identity and false façade.

The Masks of Transiency: From Ego-Self to Persona-Shadow

We need to be aware that when machines become as-if humans, humans become *as-if personalities*. The as-if personality, a concept initially formulated by the psychoanalyst Helen Deutsch more than 70 years ago,[209] belongs to the spectrum of borderline personality disorders. It is essentially an identity disorder. Sherwood and Cohen aptly define as-if pathology as "imitative, a way of life built on an endless series of transient identifications"[210] that replace each other, instead of being integrated into a stable sense of identity. The as-if personality has an as-if developed ego that looks well-functioning but suppresses feelings of weakness and avoids conflict with the environment, adjusting to any demand the

---

[208] Ed Pilkington, Death of 13-year-old prompts cyberbullying test case, *Guardian*, June 17, 2008, http://www.guardian.co.uk/world/2008/jun/17/usa.news.

[209] Helen Deutsch, Some forms of emotional disturbance and their relationship to schizophrenia, *Psychoanalytic Quarterly*, (1942) 11:301-321.

[210] Vance R. Sherwood and Charles P. Cohen, *Psychotherapy of the Quiet Borderline Patient: The As-If Personality Revisited*, London, England, Jason Aronson, 1994, xiii.

situation requires, changing colors like a chameleon. In the as-if personality, relatedness is not authentic, but may seem as if it is. Pseudoaffectivity and the capacity to wear *personae* according to the roles required by the momentary situation reflect an avoidance of intimacy. The as-if person does not really feel alive.

In *Eyes Wide Shut,*[211] Stanley Kubrick portrays the as-if personality in modern society, lacking a true sense of identity behind the seemingly well-adjusted *persona*. For Bill (played by Tom Cruise), the professional persona as Dr. Harford serves as shield against feeling and relatedness. He finds his way to a mysterious masked party. When unmasked, uncovered, his life is "redeemed" by the sacrifice of a "mysterious woman."[212] Later found dead, she is possibly identical to a drug addict he had previously saved. Toward the end of the film Ziegler, the party host, tells Bill, "Suppose I said all of that was staged, that it was a kind of charade? That it was fake?" Bill, who by then through his night journey has become more sincere, replies, "What kind of fucking charade ends with somebody turning up dead?" Ziegler retorts in a way that discloses *the evil of as-if*, because there is *never anyone responsible*. He says, "Listen, Bill. Nobody killed anybody. Someone died. It happens all the time. Life goes on. It always does until it doesn't."[213] Bill's wife, Alice (Nicole Kidman), offers him a mirror for reflection when, upon returning home, he enters the bedroom and to his dismay sees that she has replaced him with the mask he had worn and tried to hide from her; the mask is now lying next to her on a pillow "bath[ing] in moonlight."[214] She has replaced his persona of indifference with the mask behind which Bill could not hide.

The Nazis were *the* Masters of Deception, but we are perhaps not aware enough of the extent to which a world in which images

---

[211] *Eyes Wide Shut*, 1999. Screenplay by Stanley Kubrick and Frederic Raphael; Directed by Stanley Kubrick. Based on the novel by Arthur Schnitzler, Dream Story, in *Eyes Wide Shut/Dream Story*, NY, Warner Books, 1999.

[212] Kubrick, 109.

[213] Kubrick, 156-159.

[214] Kubrick, 159.

have detached from the material world, the map from the territory, may become a warring world of illusions. The perversity of deception in the service of evil compounded into the dust of the extermination camps, but on the way Theresienstadt served as a model of deception.

By exaggerated trust in collective consciousness, so called "prominenten" German Jews were lured to "Theresienbad," not realizing their choice of room at the lake or by the city square was part of the Theresienstadt make-believe. In reality less than one-sixth of the more than 140,000 Jews who passed through survived, and less than one-tenth of the children—one of those is the young woman portrayed on the slide, in what to me is a shattering photo-graph, as she departs on the transport from Theresienstadt to Auschwitz.

By the make-belief aspect of collective consciousness, the Jews were lured into the shadowy abyss of destruction. And when Helen Deutsch, who had left Vienna for the United States in 1935, wrote her 1942 paper "Some Forms of Emotional Disturbance and Their Relationship to Schizophrenia," introducing the concept of the as-if personality, the poet Leo Strauss wrote, in Theresienstadt, what in its subtle simplicity to me is one of the most spectacular poems, Als-Ob, *As-If.* I hope my English translation does not detract too much from the original:[215]

> I know a little tiny town
> a city just so neat
> I call it not by name
>   but call the town As-if
>
> Not everyone may enter
> into this special place
> you have to be selected

---

[215] Erel Shalit, *Requiem: A Tale of Exile and Return,* Sheridan, WY: Fisher King Press, 2010, 9-10.

from among the As-if race
And there they live their life
as-if a life to live
enjoying every rumour

As-if the truth it were
You lie down on the floor
as-if it was a bed
and think about your loved one
    as if she weren't yet dead

One bears the heavy fate
as-if without a sorrow
and talks about the future
    as if there was – tomorrow

In a world in which we are replaced by machines and bombarded by images, where patience, commitment, depth, and interiority are exchanged for speed, transformation, triviality, and appearance; in a world in which the ego is not only willing to wear any mask that seems to befit the moment, but in which the shadow permeates the fibers of the persona in the absence of a regulating ego, the transient personality is a more well-adapted visitor than the suffering wanderer searching the soul's path.

Insofar as we think of the ego as captain of the ship, we must be aware that the ship we are sailing through life is kept upright and in balance, and its orientation is maintained by the gyroscope or gyrocompass, in line with the internal axis of the earth; i.e., something beyond one's individual soul. Without the ego as captain and Self as gyroscope, the ship will crisscross the ocean, doomed to erratically and ghostlikely sail the worldwide waters like a Flying Dutchman.

When the ego is rooted internally in its own center, the Self, the archetype of meaning, can permeate the fibers of the personality and imbue its organs, such as the persona, the mask of appearance and social adjustment, with meaning and symbolic life.

CHAPTER

Recollection and Recollectivization:
The Transient Personality in Search of Memory

On Memory

Memory is crucial to human civilization and an important aspect of the psychization process, transforming our instinct or reflexes through reflection. While Pavlov claimed all types of psychological activity are based on reflexes, while not completely true, it is true that we lack adequate psychological activity based on reflection.

Psychoanalysis began with hysteria, dreams, and memory, the voice that psychoanalysis, the science of imagination, called for, in order to exclaim, to call out, loud and clear, that there are sexual desires and erotic phantasies, repressed, seemingly forgotten, hiding behind the *prudefemme*, behind the strait-laced appearance of prissy prudishness.

The *dream*, insofar as we have not yet dismissed Freud's theory[216], served as the secret repository for the wishes yet to be fulfilled, if we remember to find our way there through the hidden labyrinth. And the *screen memory* was there to tell us that what we

---

[216] 6 May 1856 – 23 September 1939.

think we remember is not the real thing, which deviously hides behind the deceptive screen of triviality.

In Jungian therapy and analysis, we often begin by going back in time to re-*member*, to re-*call*, and to re-*collect*: In our search for lost times, *a la recherche du temps perdu*, we *remember*, gather those scattered fragments, that if merely left behind in bygone times, leave our souls scarred by repression, their sense of meaning obliterated by denial.

We *recall* by attending to those voices that echo back at us from within and that hopefully called us from without to come home for dinner, and that reverberate to us from our ancestral past, and, if we are sensitive enough, we hear the voices that call upon us from within the womb of the future.

We *recollect* by bringing our *selves* back home or by bringing *ourselves* back home to the Self—the "act of self-recollection," says Jung, the "gathering together of what is scattered," he contends, indicates the integration and the "humanization of the self."[217]

So, when Carl Gustav Jung[218] walks down those unfamiliar stairs in the dream of his now familiar multistoried house, descending to the cave cut in the rock, digging out the scattered bones and broken pottery, the very old human skulls call upon him from under the dust. "Then I awoke," writes Jung. That is, those half disintegrating old skulls at the bottom of his dream woke him up, made him aware that "archetype is memory," memories that extend beyond the personal experiences of one's individual life, reminding us to stir up the ancestors from the dust, to let them clear their rusty voices so that we can hear the stories of the hidden treasures and so that the ghosts that otherwise lose their shape in the mud can regain their contours in complex dress.

Thinking of dust, when my son-in-law asked my 2 1/2-year-old grandson what I do for a living, he said "grandfather hides dust." I will leave it up to you to decide how to interpret that, but

---

[217] C.G. Jung, Transformation symbolism in the mass, *Psychology and Religion: West and East*, CW 11, par. 400.
[218] 26 July 1875 – 6 June 1961.

in reference to Jung, may it be that when we call out for the soul, upon our return, that we "shake the dust of all the lands from our feet?" Or that in the dust the souls hidden in matter shake off their earthly garb and rise to the clouds that bring the moisture of life back to earth?

There is *flavor* to memory, just as images flourish in our feeling. There is a smell and a taste to recollection, a sweetness and sentiments that arise when we look at a photo album, e.g., recalling the grains between the toes on childhood's sandy shores. The whiff of a certain woodsy scent that reminds us of our grandfather's aftershave, his scratchy cheeks, his quiet care.

For Marcel Proust[219], a spoonful of the tea and the taste of the madeleine brings him back to the memory of the old gray house upon the street, and "in that moment all the flowers in our garden and in [the] park, and the water-lilies on the Vivonne and the good folk of the village and their little dwellings and the parish church and the whole of Combray and of its surroundings, taking their proper shapes and growing solid, sprang into being, town and gardens alike, all from my cup of tea."

The taste on the tongue and the songs in our soul and the bodies that tenderly touch each other's hearts, may all be hastily swept away as we swipe the touchscreen in a futile attempt to look at the ceaseless stream of digitized photos, no longer identified by name and place and date, but numerated into infinity; "*the many*" has replaced "*the one*," and we hardly return to pay respect, to spect-again, to re-*spect* by looking again, perhaps having forgotten that the soul resides in what we respect and revisit, in the sense of looking again, in the *exploration* of the *experience*.

Memory is crucial to the human condition. Often have I heard those who returned from hell say that you might survive the Shoah, but you cannot survive Holocaust *denial*.

A patient of mine who had emerged from years of hellish suffering needed to tell and hear and tell the stories of hell in order

---

[219] 10 July 1871 – 18 November 1922.

not to forget her suffering, so that she would remember the sense of life's meaning, imprinted in years of pain and struggle.

Now, we are told, the new cure for post-trauma is memory deletion, the forgetting pill that erases painful memories—forever; at least I think it is forever, if my memory doesn't fail me. No longer shall the memory of the horrors of war and rape, of terror, murder or abuse haunt the suffering soul. As the author of that unforgettable article says, "In the very near future, the act of remembering will become a choice."[220]

In parenthesis, only a few weeks ago it was revealed that the author of that article inserted faked quotes in his book *Imagine*, trusting his readers to have lost enough of their memory not to question his lack of integrity and perhaps confusing imagination with imitation—one of the sins of the postmodern era.

Were I to suffer from severe trauma, perhaps the forgetting pill *would* be my treatment of choice. I may have rushed into the promising embrace of painless forgetting, but I remain skeptical and concerned whenever I hear the promises of shadowless utopia, perhaps unduly having a perverted preference for dystopia.

There are memories that are unbearable, memories whose weight is so heavy that we cannot bear it, memories that humiliatingly force us down on our knees. Yet, artificially erasing them may carry the even heavier weight of the loss of mind and psyche. The Norse god, or giant, *Mimir*, the wise rememberer, tells us that *The Well of Knowledge and Wisdom* is guarded by *Memory*. Without memory the voice of the ancestors, the accumulated knowledge from yesterday, and the depth of wisdom from yesteryear merely evaporate.

Neither does the ease of access in today's world to a wealth of information automatically make us richer in knowledge, nor does it necessarily bring us closer to the treasures of wisdom. Rather, "the dream of full access to human knowledge amounts to a state of ignorance," as Daniel Matiuk says, and we arrive at the

---

[220] Jonah Lehrer, http://www.wired.com/magazine/2012/02/ff_forgettingpill/.

poverty that stems, according to Walter Benjamin, "from the divorce of experience and tradition."[221]

Memory is the grand achievement of psychization. Without it, there can be no representation, which is the pillar of human civilization, extracting us from the condition in which we merely do and act, without the *mental* experience of the *actual* experience. Without recognition and representation, both relying on our capacity to recollect, we live the unreflected life.

Is it not for memory's sake that the baby spends most of its time in a state of dreaming, just as the old person often spends time recalling and remembering, unless dementia has cruelly stolen the recollection from him or her? In a cycle that repeats itself daily, life transforms itself on an axis from the infant's unmediated experience to the old person's recall and recollection of remembered experience.

Thus, memory and recognition, continuity, constancy and reliability of memory ripen and develop in contrast to a condition of erratic fragments of unrelatedness.

Memory and the Mass Man

The constant onslaught of stimuli on the city dweller's psyche requires the individual to adapt by *disengagement*, claims Benjamin[222]—who would have been 127 in a few months of this writing if, having safely crossed the French-Spanish border but fearing repatriation into Nazi hands by the Franco Regime, had he not committed suicide, or died some other way, at the very "center of misfortune," as Hannah Arendt said, on September 25, 1940. *Disengagement* and alienation impair the capacity to appreciate poetry; in fact, Benjamin considered urbanization to be the internalization of forgetting.

---

[221] Daniel Matiuk, in Mätiuk, D. and Yoran, N., 'The Humanities' Inner Wound: The New Poverty of Experience and the Dilemma of Waste,' in *The Future of the Humanities 2*, p. 4f.
[222] 15 July 1892 – 26 September 1940.

And, we may add, an overdose of defenses is suicidal to the soul, causing it to freeze or go into exile, even if there may sometimes be no alternative in the shadow of Holocaust and Deceit, of Fascism and Fragmentation, of Totalitarianism and Atomization.

From the perspective of Analytical Psychology, Jung and Erich Neumann dealt prominently with the soul and ethics of modern man and mass-mindedness.

Jung writes in a letter in 1947 that "... collective systems ... have a destructive effect on human relationships. ... [T]otalitarian States ... undermine personal relationships through fear and mistrust, the result being an atomized mass in which the human psyche is completely stifled."[223]

We may add that quantification, for instance by demographics and statistics, is our way to conceive a recognizable persona of the group in its lack of the qualifying character we find in the face of an individual.

The atomized mass, however, is not solely a consequence of the Totalitarian Regime or the Totalitarian Mind: rather, the unintegrated mass-man in our soul lacks integrity, the capacity to hold out against fear and mistrust and, therefore, may all too willingly wear the uniform, the one and single form or shape of manifest or latent totalitarianism.

Neumann describes how the detachment from the original group and from the unconscious has, on the one hand, degenerated into overspecialization and on the other hand "exalt[ed] the mass as a conglomeration of unrelated individuals."[224] The homogenous group making up the tribe, clan, or the village has been exchanged for the mass units of city, office, or factory, says Neumann.[225]

And now, 70 years later, the mass units of the city and the factory have been replaced by the megaunits of *Faceless Friends on Facebook* and the Avatars of *Second Life*. The quantitative limitations of so-called reality have been broken up by boundary-

---

[223] C.G. Jung, *Letters*, Vol. 1, p. 472.
[224] Erich Neumann, Mass man and the phenomena of recollectivization, *The Origins and History of Consciousness*, p. 436.
[225] Ibid.

less virtuality—25 million people have watched a perhaps cute but trivial depiction of the birth of a baby avatar on Second Life.

Neumann says that what we get when consciousness is overpowered and possessed by the unconscious mass component, is the mass man, "as manifested in the mass epidemics of recollectivization."[226] We might deceive ourselves that we have attained historical heights of individual development, but the post-modern condition reflects, rather, disorientation, fragmentation, superficiality, and atomization.

While in general we partake less and less in external reality, we can witness the process of enantiodromia, when we see the young who leave their lonely communication chambers and take to the streets and the city squares. The fact that the spirit of youth does not ascend to the corridors of power is a political cruelty and a psychological tragedy.

Mass man resides within; mass-mindedness is a state of mind, and the collective shadow is most prominent in the disengaged individual. While the negative element, as Neumann says, essentially has "a meaningful place as decomposition and death, as chaos and *prima materia*, ... in a fragmented psyche ... it becomes a cancer and a nihilistic danger."[227] Neumann notices that it is the cleavage between ego consciousness and the unconscious that enables the regression to the mass-man in us, and he points out that the mass "is the decay of a more complex unit not into a more primitive unit but into a centerless agglomeration."[228]

A "centerless agglomeration" is the antithesis of centroversion and of a centered, focal entity, which we may term a, or *the,* Self. Centerlessness is a characteristic of the postmodern condition, thus perhaps both a consequence of historic regression, as well as a catalyst for a process of disengagement, fragmentation, and transiency. The pseudounity composed by a "herd of atomized individuals," says Neumann, "is sheer illusion."[229]

---

[226] Ibid., p. 439.
[227] Ibid., p. 440.
[228] Ibid., p. 441.
[229] Ibid., p. 442.

Could that be more true than today! While Neumann refers to transient mass-phenomena, such as mass-meetings, the very same holds water in a world where the boundaries between the virtual and the real are blurred, in which the meeting place is a Facebook page, where *friends* are no longer individuals-in-relation but numbered in the hundreds, and *like* is not a real value and a feeling, but a button and a rating.

While we desire transparency in management and politics, not only Wikileaks; the boundaries of privacy are porous, broken through in the dual reality world where practically every event is photographed and posted on the web.

Once upon a time we claimed that an event that was not related to seemingly did not take place, since it is not registered in consciousness, in conscious memory, and therefore has a tendency to repeat itself compulsively. This is the way complexes work—they repeat themselves compulsively, detracting energy from consciousness, rather than being integrated into consciousness.

This view is quickly becoming obsolete, better forgotten. Now, the event did not take place if it is not registered visually, digitally, and preferably posted on the worldwide fairy web.

One of the characteristics of this collective reservoir is the lack of differentiation—a stage of consciousness that has not yet been fully reached; gold and trash mix, and it does require the wisdom of the alchemist to extract the gold from the garbage.

Processes that once were thought of as internal, pertaining to reflection and interiority, are now increasingly exteriorized, and we all become addicted to speed and impatience, to quantity and masses, to association and fragmentation.

I am concerned that in an era of recollectivization, the alchemist, the soul seeker, is a rare species, and the undifferentiated ability to associate, to move between site and post, a more common trait. The patient and determined persistence of the alchemist is replaced by the uncommitted and impatient movement of the transient personality.

PART IV
Restorying the Self

CHAPTER

15

Self, Story, and Transiency in the Postmodern Condition

In the Beginning

In the story of Genesis, we are told that God created light by raising his Voice, speaking the words "Let there be light."

Perhaps the light was needed in order to tell the story, to let the story of transcendence and divinity light up and be illuminated, to manifest in this world. According to the ancient Kabbalistic *Book of Creation*, or of *Formation*, ספר היצירה, the world was, straightforwardly, created by story and by storytelling. And in the story of Rabbi Nachman of Breslov God made man because he loves stories—sometimes we might wonder what tales and fables of man express God's great love of stories...

While *silence* pertains to sheer interiority, the word and the *story* refer to light, consciousness and manifestation. Silence and story are siblings, like image and dialogue, shadow and Self, awe and wisdom.

The story brings form to the void, light to the darkness, substance to nothingness, flesh to the emptiness (Genesis 1:2). Myths are the stories that give the gods a mouth to speak, and fate is the often mischievous story woven in the mind of the gods. The story brings the archetypal idea into the complexity of reality. The poets receive their authority and drink their mother's milk from Mnemosyne, goddess of memory of what has been, mother of poetry and tragedy, of history and of the hymns.

> And the story brings life to the unconscious. As Jung says: Life is (the) story of the self-realization of the un- conscious. Everything in the unconscious seeks outward manifestation, and the personality too desires to evolve out of its unconscious conditions to experience itself as a whole.[230]

The Story of the Self's Manifestation

Life is the *story*, says Jung. "Without a story of your own, you haven't got a life of your own," says Laurens Van der Post. And James Hillman states that the psyche pertains to "getting the subjectivity out of it, so the story, the image takes over."[231] "One integrates life as story," he says, "because one has stories in the back of the mind," i.e., in one's unconscious, which then serve, Hillman continues, "as containers for organizing events into meaningful experiences."[232] The story, he says, might then be "an anecdote of a wider truth." Just as the Self may be understood as the spark of the divine in the human soul, the story pertains to the Self's manifestation in life: to count and to recount, to account and be accountable, pertain to the conscious life that Jung speaks about.

In the Grimm tale about the Devil's three golden hairs, the protagonist makes the journey three times over:

First, the easy journey of the divine child, the original, primary Self, equipped with good fortune, who with a little help

---

[230] Jung, C.G. (1962), *Memories, Dreams, Reflections*. New York, NY: Pantheon Books, p. 3.
[231] Cf. Dick Russell, *The life and ideas of James Hillman* (loc 279).
[232] James Hillman, A Note on Story, in *Loose Ends*, p. 1.

from his friends and foes arrives at the king's castle, totally unaware of the dangers on his road.

Then, he departs on his journey to meet the devil in hell, still optimistic, childishly naïve, and greatly inflated, yet heroically open for adventure. Guided by powers and capabilities not his own, he claims to know everything but provides no answers. However, he is now ready to be asked the questions of life, to encounter the problems of the world:

- ✖ Why is there no water in the well in which once wine would flow?
- ✖ Why does the tree which once bore golden apples no longer bear fruit, not even put forth leaves?
- ✖ And why does the ferryman have to eternally row back and forth between the riverbanks, never set free?

And finally, he needs to grow up, to mature. On his journey back from being a humble ant in the folds of the Great Mother, in the devil's grandmother's gown, he has learned not only the questions of life, but also the answers given him by the devil, that instigator to the conscious life:

- ✖ First, how the wellsprings of wisdom of the earth, the domain of the Mother, cannot flow, neither with water nor with wine, neither with life nor with spirit, if the toad has been confined to sit under the stone, if the feminine has been subdued and repressed.
- ✖ Secondly, how the fruits of wisdom and love, the golden apples of knowing Eros, dwindle if we deny the shadow, if we don't pay attention to the rat that is gnawing persistently at the roots of the tree.
- ✖ And thirdly, how the movement between the opposites loses its anchor in the Self when we are stuck in the repetition-compulsion of a complex that traps us in meaningless, perpetual rowing back and forth between the river banks, never set free.

And yet, were there no trouble in the kingdom, there would be no story; no story that would ask us, as does the old woman in the forest ask the young boy in our story, from where do you come and where do you go?

That is perhaps why Jung, in his essay on the stages of life, written in the early 1930s, stresses the importance of *problem* as regards life's stages, and Rilke expresses the crucial importance of problems for the sake of recognizing oneself, that is, one's sense of identity. In *Testamente*, he writes:

> Ever since I was a child I believe I have prayed for *my* problem alone, that I might be granted mine and not by mistake that of the carpenter, or the coachman, or the soldier, because I want to be able to recognize myself in my problem.[233]

It is in *my* problem, and in my *problem* that I recognize myself. If a person turns away from the suffering that a psychological conflict entails, he or she will, instead, suffer from a neurosis, says Jung.[234] Pathology ensues when there is an avoidance of problem, when everything is assumed to be without problems—which commonly leads to problems for others, for instance when children have to carry their parents' split-off shadows. Without problems I become flat, a flat surface without personality, without an inner voice—the voice of myself, of who I am and conceive myself to be, as well as the Voice of the Self. Without that voice, I become an impersonal mask, a faceless persona, an un-individuated part of the collective, unidentifiable, without individual colors and shapes and complexity.

My *problems* make me into the unique individual that I am. While there is nothing romantic about pathology, crucial meaning can be found in suffering, and in order to find the depth of meaning, I must often be ready to suffer.

---

[233] R. M. Rilke, Testament in *The Inner Sky: Poems, Notes, Dreams*, Daimon Searls, Trans. (Boston, MA: David R. Godine, 2010) 85.
[234] CW 18, par. 383.

Without complexes that engage the ego, there is neither complexity nor embrace of the other; whether the other *person* or the *shadow* or the transcendent *other*—the word complex, remember, comes from *complectein*, to embrace.

This, I believe, is the reason that Jung emphasizes *problem* so strongly in his brief essay on "The Stages of Life,"[235] because the problem-free life entails no challenges, no call for one's individual path; no conflict or *complectein*, neither complex nor embrace.

How much *problem* is not caused by our neurotic defenses, erected in our attempt to avoid being burdened by our problems and challenges! While we all have to travel the journey through life, the neurotic person is a reluctant traveler, who, as Otto Rank said, refrains from taking the loan, that is, the load of life, because one day we have to pay it back.

The Epic Side of Truth

In his exceptional essay "The Storyteller" from 1936, Walter Benjamin tells us that "the art of storytelling is coming to an end." The reason, he says, is because *the communicability of experience is decreasing*. Benjamin refers to World War I, how the men returning from the battlefields had grown silent. He writes:

> Beginning with the First World War, a process became apparent which continues to this day. Wasn't it noticeable at the end of the war that men who returned from the battlefield had grown silent—not richer but poorer in communicable experience? ... [N]ever has experience been more belied than strategic experience was belied by tactical warfare, economic experience by inflation, bodily experience by mechanical warfare, moral experience by those in power. A generation that had gone to school on horse-drawn streetcars now stood under the open sky in a landscape where nothing remained unchanged but the clouds, and, beneath

---

[235] CW 8

those clouds, in a force field of destructive torrents and explosions, the tiny, fragile human body.[236]

The body freezes along with the spirit and soul into a silence that is impenetrable, into a shadowy realm of living death. The horrors prevent communicability of experience, as we know from Holocaust survivors.

Benjamin then elaborates on seeking counsel, when, he says, it:

is less an answer to a question than a proposal concerning the continuation of a story which is in the process of unfolding. To seek this counsel, one would first have to be able to tell the story. ... Counsel woven into the fabric of real life [*gelebten* Lebens] is wisdom. The art of storytelling is nearing its end because the epic side of truth—wisdom—is dying out.[237]

When we seek counsel, we thus search for the continuation of our story. When the unfolding of our story is blocked, we fall into depression; when the story becomes erratic, we are struck by anxiety.

Benjamin calls wisdom *the epic side of truth*. That is, to him, wisdom is the long path of pursuit and struggle, the hero's journey and ordeal. The story, we might say, is the hero's journey—the hero, of whatever gender, who departs from the safety and security of home, who tears the fabric of the familiar, who carries the pain and the guilt and the irresolvable conflicts on shaky legs and weak shoulders, who searches for the treasures of soul and meaning that he or she might hopefully find in the scary lands of the night and the dark, the hidden riches in the dream, returning never in triumph, but perhaps in silence, carrying the paragons of the beyond in the sadness of tragedy that is present whenever we break the bonds of home.

---

[236] Walter Benjamin, 'The Storyteller,' in *Selected Writings, vol. 3*, p. 143f.
[237] Ibid., p. 145f.

Analysis is a way of seeking counsel, of letting one's story unfold. But we hardly seek counsel these days, because often our story does not unfold by pursuit and ordeal. We often refrain from the epical journey because, as expressed so succinctly by Paul Valéry, "Modern man no longer works at what cannot be abbreviated."[238]

From Benjamin's perspective, and I could not agree with him more as regards our postmodern condition, the antithesis of *the story* is *information as a form of communication*. He writes, "Every morning brings us news from across the globe, yet we are poor in noteworthy stories." By means of cyberspace, we are fed external images and items of information around the clock. While "the intelligence of earlier centuries ... was inclined to borrow from the miraculous," says Benjamin, "information must absolutely sound plausible."[239] Adding to Benjamin, Susan Sontag writes:

> There is an essential—as I see it—distinction between *stories*, on the one hand, which have, as their goal, an end, completeness, closure, and, on the other hand, *information*, which is always, by definition, partial, incomplete, fragmentary.
> This parallels the contrasting narrative models proposed by *literature* and by *television*.
> Literature tells stories. Television gives information.[240]

Story and Information

The *story*, we might thus say, is touched by the miraculous, anchored in the layers of epic wisdom, while *information* is embedded in the deceptive simplicity of plausibility. Stories and storytelling dwell in the realm of the objective psyche, molded by

---

[238] Quoted by Benjamin, ibid. p. 164. From Paul Valéry, *Degas, Manet, Morisot*, trans. (David Paul. Princeton: PUP, 1989) 173.

[239] Ibid., p. 147.

[240] Susan Sontag. *At the same time: Essays and speeches.* Edited by Paolo Dilonardo and Anne Jump, with a foreword by David Rieff, (New York, NY: Farrar Strauss Giroux, 2007) 221.

archetypal images. The story emerges from the wellsprings of archetypal images that circle the world of psyche. Plausible information will fit the confines of convention, of collective consciousness, while the narrative of the story dips into the lakes of wisdom, into the Self as archetype of meaning and as the psyche's faculty of image- and symbol-formation, to extract the essence of the Self's products. James Hillman says, "The difference between ego and psyche isn't only theoretical; it's in how you tell a story. It's in getting the subjectivity out of it, so the story, the image takes over,"[241] and the story becomes "an anecdote of a wider truth."[242]

However, in today's world we have neither the mental time nor the reflective space to attend to the Self and the world of symbols and images. Steven Spielberg says, "People have forgotten how to tell a story. Stories don't have a middle or an end anymore. They usually have a beginning that never stops beginning." [243]

Impatiently we strive for abbreviation, as Paul Valéry says, for the blurb, the short message service, the SMS text, until we empty our story into the communication of a maximum of 140 characters on Twitter; which means a chirp or a chatter. Twitter creator Jack Dorsey confesses that "twitter" correctly defines the product as "a short burst of inconsequential information."

That certainly is an honest revelation from a trustworthy source: 500 million registered users provide 340 million daily bursts of unimportant information—a short sentence for Twitter, a giant waste for mankind.

Twitter and Facebook are manifestations of our need to tell *a* or *our* story, but it is often fragmented, simplified, and abbreviated into the language of commercials. The story has been reduced to information, the collective unconscious replaced by a collective of unconsciousness. But if there is no story—and of

---

[241] Inter Views, 100.

[242] Case History: Evolution or Revelation? in Jeffrey Zeig, *The Evolution of Psychotherapy…,* (see Russell, life and ideas of JH, p. xxii, note 27). 287.

[243] http://oddernod.com/post/8972021458/people-have-forgotten-how-to-tell-a-story

course we still have stories and a need for stories!—I am only trying to pose one extreme against another in order to illuminate the dilemma and the current processes: If there is no story, what happens to the author, who supposedly has some say, even if little authority, over the life he or she is supposed to be authoring?

Well, in 1967, the French philosopher Roland Barthes announces that the author is dead! So, what happens to the story, when the author is dead? I am not asking what happens to the story after the death of its author, because if the story is good enough, it may serve as the legacy of a life fully lived or fully authored—no, I wonder about the story written by a dead, by a nonexistent author, an author who cannot be accounted for, in fact, a story without an author?

When the author of one's life is deconstructed, the story becomes erratic—neither author nor story. Without our story, we are not fully awake, or as Rabbi Nachman of Breslov says, "While stories are said to help you sleep, they are meant to wake you up!"

When Scheherazade tells her thousand and one nights of stories, the feminine remains alive, and there is no execution of women. Without stories to wake up from, men tend all too easily to become all too literal, sometimes hardening into fundaments and principles.

As Hillman points out, "The source of images—dream-images, fantasy-images, poetic-images—is the self-generative activity of the soul itself."[244] Stories deliteralize, just as images give volume to the soul.

The Transient Personality

In what way have we lost the story today?

Perhaps we have lost the epic side of truth, the wisdom that requires the epic journey, the long path of pursuit and struggle, the actual encounter with the tree that bears no fruit; we do not drink

---

[244] James Hillman, *Archetypal Psychology: A Brief Account,* (Dallas, TX: Spring Publications, 1981) 6.

from the spring without water, and we refrain from the hard and repetitious work of rowing across the banks of the river, stuck, rather, in permanent transiency.

As a consequence, we receive little or no wisdom from the devil. Rather, without the epic journey, the wisdom of the devil turns into evil; "The reason for evil in the world is that people are not able to tell their stories," [245] says Jung. The reverse is likely to be equally true: because of the evil in the world, people have a difficulty telling their stories.

The often-painful journey that expresses the story told by the Self has been replaced by transiency, the postmodern shadow of what Freud describes in "On Transience." Let me initially illustrate the phenomenon of transiency by using the GPS, the satellite navigating system as metaphor. (Nowadays when ideas and inventions take the shape of computerized images rather than mechanical machines, image, metaphor, and application tend to merge.)

When the hero departs for his journey, to retrieve something that has been lost to consciousness, to the realm of the ego, he or she needs to recognize the call and to be attentive to the helpful creatures along the way and particularly, to be equipped with ego strengths, such as endurance, determination, and an inner sense of calling, vocation, a voice that leads the way and knows the goal.

The navigating system reflects how we revoke ego-control. We are easily seduced by the visual depictions on the screen, the 3D images that often seem more attractive than reality. The machine becomes the leader on which I rely. Even if I know a different way, which I normally would rely on, I become hesitant and insecure if it deviates from the machine, computer, from the robot. It really may not be very important on the road, but when a doctor hands over his or her knowledge, experience and intuition to rely solely on biopsy, CT, and blood tests—all valuable as diagnostic tools in the right person's hands—when the doctor becomes a servant to the machine, he or she has lost the anchor in the divine physician, the archetype of the physician.

---

[245] Letters Vol. 2, https://www.goodreads.com/quotes/439317-the-reason-for-evil-in-the-world-is-that-people.

With the navigating system, we need not see the whole picture, to map our route from a point of departure to destination, the place for which I am bound, i.e., psychologically, my destiny. I am guided as if blind, I need not make any choice of route or see the meaning to be found at the crossroads; the road is no longer My Way, my story. Unlike the tale of the devil's golden hairs, I do not stand in front of the woman in the hut in the dark forest who asks the Godlike question of destiny, "Whence do you come, and whither are you going?" I follow the voice that calculates and thinks for me, and I follow the fragmentary pieces of the map as I am led along, without an overview, calculation and voice of my own.

You might think that I exaggerate about the GPS, but a study from 2006 shows the difference in brain development between cab and bus drivers in London. The latter drive their routine line, the former need to learn and be alert about the web of streets, how to go from one place to another in that complicated web, which leads to a significant difference in the development of the hypothalamus. So, no consequences of being blindly led by the machine?

The story dips into the archetypal depths, it is a journey, has a center, a problem, a kernel, a memory, a point of departure and a destination. In contrast, in postmodern transiency, which is not to be confused with Freud's emphasis on the *value* of the transient, we lack a comprehensive view. The anchor of experience in Self and interiority has been replaced by externally generated information.

While Freud speaks to Rilke, in 1913 (*On Transience*)[246], about the ability to mourn what will be lost, when the poet finds it difficult to enjoy beauty because it is transient, the Transient Personality avoids the sense of loss and the need to mourn the loss by never staying in one place, or, if he or she does appear to be present, it is an inauthentic presence, a "photographic presence," like the colorful computerized images for instance of the GPS.

I believe that Umberto Eco speaks of the Transient Personality when he says, "The completely real becomes the completely fake."

---

[246] Freud, Complete Psychological Works, Vol. XIV, pp. 305-307.

The Meaning and the Message

Freud emphasizes how treatment begins with the patient telling "the whole story of his life and illness,"[247] and in *Memories, Dreams, Reflections*, Jung says, rather characteristically:

> Clinical diagnoses are important, since they give the doctor a certain orientation; but they do not help the patient. The crucial thing is the story. For it alone shows the human background and the human suffering, and only at that point can the doctor's therapy begin to operate. [248]

The story pertains to life, illness and suffering. This is the story, one's own story, neither fiction nor fantasy, nor photographic or so-called factual reality, but a subjective documentary, which combines the events we record in memory, called anamnesis, and the inner world of fears and wishes, joy, and despair, the interiority of images and imagination as well as the external world of object relations.

Without *problem* there is no story and no vital relationship to the Self, the archetype of meaning, the sense of meaningful interiority. And how much suffering does not emanate from the neurotically unlived life! How burned-out do we not become when we invest most of our resources to extinguish the fire of life! To be the author of one's life means to live what Jung calls the conscious life, which requires that we know ourselves. In his speech of defense at his trial of heresy, Socrates succinctly states that "the unexamined life is not worth living."

The meaning and the message of the story are more important than symptoms and diagnoses. That is, the story is made up of meaning and message, anchored in the Self as archetype of meaning and as daemonic guidance.

---

[247] Dora, 1905.
[248] Jung, *Memories, Dreams, Reflections, 124.*

CHAPTER

16

Image, Meaning, and the Healing Self

Does the flow of our imagination come to a standstill when we are depressed, or is depression a result of obstructed imagination? Does depression reflect a condition in which the psyche has frozen in the grips of a devastating image? Then is feeling, relating, value, any kind of emotion, however sad and painful, an antidote that enables movement out of depressive stillness? Jung imagined his unstable emotions as containers, as vessels on the verge of breaking. His brilliant insight told him that the healing images needed to be extracted from his troubled emotions. Were they to be left hidden in the emotions, either he might "have been torn to pieces," or he might have split them off, which would also have led to being "ultimately destroyed by them anyhow."[249] Enlivening the images, to retrieve them from their hiding place in the shadow, to carve them out of the stone or to shape them out of the clay, to speak with them and let them talk, is the core of the healing process.

And in *Nineteen Eighty-Four*, Orwell writes, "Tragedy, he perceived, belonged to the ancient time, to a time when there were

---

[249] Jung, 1961/1965, 177.

still privacy, love, and friendship."[250] "But," as Dürrenmatt explains, "the tragic is still possible. ... We can bring it forth as a frightening moment, as an abyss that opens suddenly."[251] Today, our survival may depend on our capacity to suffer depression, states of anhedonia, guilt, and grief and the painful memories of the past, as well as the anxiety of the future. If these characterize the survivor syndrome, they may be necessary characteristics for our survival. As Carl Sagan has said, "The price we pay for anticipation of the future is anxiety about it. Foretelling disaster is probably not much fun; ... [but] the benefit of foreseeing catastrophe is the ability to take steps to avoid it."[252] Dystopia, which in many ways is primarily characterized by the lack of Eros, is better equipped to scrutinize the shadows that utopia carefully avoids.

The experience of storytelling, time out of time, forms the basis for a safe existence, the foundation on which the ego can develop. In order to be able to cope with the doubts and insecurities in life, one needs to have the experience of security as a basic frame in life; then, one can live with the insecurity and doubts of the ego. We all grew up with fairy tales, nursery rhymes, and myths that transcend borders and have burrowed deeply into our psyches, into our souls as lessons from the past on how to survive, how to deal with evil, monsters, the overwhelming challenges of life. The important stories form the archetypal net that holds all of us together in our ever-expanding groups from generation to generation. Once upon a time is out of time but instead located in imaginary time, time of being, outside of the ego, and time we all participate in. These are lessons in imagination, and each story takes on our own shade of color, we each step into psychic time at our own chosen location and follow the path that calls to us.

These stories are passed down both orally and in written form, expanded, changed to fit the new circumstances, but they are always there as a backbone that links the groups' members to each

---

[250] George Orwell, *Nineteen Eighty-four*, ch. 3.
[251] Dürrenmatt, p. 34.
[252] Carl Sagan *The Dragons of Eden*. Ballantine Books, New York, NY: 1978, p. 74.

other and to experience. One such group is the Jews, who may not be a race, genetically homogenous, nor merely a religious group that shares the worship of God. But they have stories—not only the story of the Jews, but the Jews as a story being told and retold, whether it is the story of Passover, to be told "as if you were there yourself," or the stories of Creation and Exodus, of the Talmud and the legends, the story of its simple folk and its great men, the story of backwardness or the story of science, of poverty and of capital, or the story of pogroms, persecution, and destruction, and of renewal. And it is the story of Hebrew tribes and how they ended many journeys and travails finally settling in their own land. These stories remain through major collective changes in the world, through wars, exiles, the rise of industrialization, modernism, and now,[253] from within the very characteristics of the postmodern condition with all its challenges, the healing story, the story of healing images continues to be written. We learn from these stories, from the past as they move into our future.

The restoration of Self, the archetype of meaning and the very antithesis of imitation, will be discussed. This will be exemplified by different methods of approaching the unconscious and the psyche at large, when writing the stories of the soul. Prominently, we shall look at Jung's active imagination, as revealed in his *Red Book* as well as the collective expression of a process of re-creation, as depicted in the Quiché Maya creation myth, *Popol Vuh*. Similarities between these two different processes of renewal—one personal the other collective—reveal a way into the deep source of life *within*, which leads to a reconnection to soul, a remaking of a worn-out organism, left in tatters by our transient world.

The cosmogonic myth has been used as a therapeutic method in many indigenous cultures. Jung's thorough study of myths and comparative religions before embarking on his own personal journey, showed him its profound value. For the Quiché Maya, the

---

[253] See Simon Schama videos on the history of the Jews.

world has been destroyed and re-created four times, a necessary cyclic phenomenon as a way to reconnect in a renewed way with the source of life. The analytic process, including alchemy, follows a similar course. Dipping back into the past as a needed precursor before moving forward, toward healing, will be explored through these two different paradigms, both by contributor, Nancy Swift Furlotti.

CHAPTER

Revisiting the Well at the Dawn of Life
Nancy Swift Furlotti

Introduction

One thing we can count on is that life is ever-changing. We have ample evidence of this from evolution, extinctions, mutations, the rise and fall of cultures, and what we experience in our own process of individuation. We never arrive at an end point but remain on a long trajectory that keeps challenging us to grow and expand our worldview. That's why we cannot say we are individuated. It is a process of continual growth and development that moves much like an upward spiral around and around, reaching points we have touched on before but at different levels of the trajectory. We would like to think this spiral generally moves upward toward greater consciousness as we individuals and the world's cultures gain more insight and understanding, but there are also times when it descends, and we seemingly have forgotten what we have learned.

The descent in alchemy is into depression, the *negredo,* where all is black and blue, and the memory of other times is inaccessible. We are forced to retreat into the desert of our being away from the lush, fruitful creatively accessible parts of ourselves.

The dry land is full of heat and sand, where nothing grows, and nourishment is limited. We wait, suffering the loss of what was once there, what we took for granted, resources we misused without thought. Thrown back on ourselves to find a way forward, waiting for some new spark of light, some knowing spirit within to guide us out of this perilous land, not to return to what we remember but to discover something new, a new way forward. We pray this spirit or soul spark will appear—tiny like a firefly grabbing our attention and then disappearing, becoming stronger as we trust and follow, allowing the psyche to lead the way.

Remembering there is such a force as a psyche—remembering its purpose, its presence, how it has shaped past worlds and has the capacity to do so again is important because psyche cannot do this alone. Its mysterious intension is intrinsically connected to each one of us as the physical being who is capable of implementing its creative design, or not. Do we move up the spiral or down? History has shown we do both. Whether we participate actively in the evolution of our cultures is up to us. Without our participation our dominance and culture recedes while the natural process of change continues on its natural course as a never-ending spiral.

I remember once hearing a physicist remark that with the inhalation of our breath we might be breathing in atoms that once made up the philosopher who was Plato. This points to the fact that life recycles itself endlessly. The disturbing feature of this statement is the implication that we are not so very important. We are made up of many parts that will be remade after we, as we know ourselves, are gone. Equally we might exhale a cell that once was part of Hitler. What a disturbing thought that is.

There is so much we don't know about the workings of the cosmos even though we presume to control our world supported by the false belief in the supremacy of rational thought, science, and technology. We are not actually the ones in control, but we may just have the ability to influence the process and the ultimate outcome for our species and the many others on this shared planet.

We homo sapiens sapiens are the result of an incredible experiment of nature that created us, animals capable of creative thinking, language, and consciousness. Yet, it is important to remember that many of us contain the hereditary marks of extinct Neanderthals. I contain 2.7 percent Neanderthal DNA. We now know that we live in a careful balance, supported by a multitude of other organisms, both extinct within our DNA and living within our gut. The template for life seems to be a reciprocity between a multitude in symbiotic relationship that makes up the one.

I am mindfully aware of how easy it is for this careful reciprocity that balances nature to go wrong. A glaring example is the specter of global climate change. This past year we experienced the reality of the hottest year on record—droughts, floods, hurricanes, typhoons, island nations buying up other lands to which they can relocate their entire populations. I live in California, where we endured five years of drought with the very real and visible experience of what that means for us locally—increased water prices, dying gardens, caring for every precious drop of water, some neighborhoods trucking in water, and the devastating effects of our local drought for the rest of the country. We are the breadbasket of the country, so this means less food, higher food prices, farmland becoming arid, farmers going out of business. And this was just the beginning. What came next was nothing less than a catastrophe and apocalypse: fires, floods, mudslides, death, destruction, a devastated community. In this list of apocalyptic catastrophes, all we were missing is the onslaught of locusts—which might yet come.

We have ignored warnings, we have gone into denial—we have not wanted to change, to believe this is possible. How could the Earth stop taking care of us? She is our Great Mother after all, isn't she? How can we psychologically understand our participation in this dramatic shift? To start, it always helps to go back in history, to past cultures, to stories and mythology, into our collective memory bank to see what happened in the past when consciousness approached a similar challenge on the grand spiral of time.

Our Western mode of belief sees life as linear, starting at a beginning point of creation and ending at some unknown apocalyptic point in the future. Being of this mindset, many frequently fear that the end-time is imminent, visited upon us from some greater force like the gods. We live with this fear while ignoring the very real prospect that it may not come from some cosmic catastrophe but instead from our own hands.

In fact, we humans are creating an environmental disaster like none other through our total disregard for a living relationship to the soul of nature, the *anima mundi*. From our worst characteristics as a species—arrogant, aggressive, self-serving, fearful animals that we are—we have wandered far away from the path that bridges soul and spirit, cosmos and nature. We, in our short-sightedness, are poisoning our environment, heating it up, causing droughts and floods, starvation, migration, loss of fertile land, reducing our rainforests that give us oxygen and fresh water, acidifying our oceans and killing them, too. All this puts pressure on our societies, and strife results. Wars break out for the remaining resources. We consider building walls to keep the inevitable flood of migrants out as more and more are forced to leave unforgiving, lifeless lands across the globe from Africa to Asia.

We were created with free will, as many creation myths point out, allowing us to make choices. Free will allows us to respect and honor our creators, those forces of many names—the archetypes that influence our behaviors—and the natural balance between all species of animals, plants, and minerals on our shared Earth—or not. We seem to have lost our respect for the Great Mother, the anima mundi that cares for us, feeds us, holds us—and has done so since we evolved out of our exclusively animal selves. With our belief that we can take from her whatever we please, in any way, and pour into her toxins that both stifle her and her other creatures, we are unwittingly choking off our source of life.

Not only have we lost our relationship to the Great Mother along the way but to the Great Spirit as well. The natural balance between the spirit world, the earthly realm, and the underworld has been forgotten. We have become sophisticated, clever, and technologically adept, seemingly rational and superior to all else on our planet, but we have forgotten the importance of balance, which includes feeling relationship. We have forgotten our creation myths, our stories, and the imperative to honor the gods, those forces greater than us. Even carrying the marks of extinction in our own DNA, we forget this can also happen to homo sapiens sapiens, leaving us not as "wise" as our name suggests.

Forgotten Worlds, Mythic Stories

Returning to history and lessons from the past, I turn to the Quiché Maya creation myth from Mesoamerica, called the *Popol Vuh*, which is a striking example of a story that lays out a template for humanity to live in balance with spirit and nature. It is a forgotten, obscure piece of writing that was almost destroyed to intentionally leave no evidence of it or its civilization's existence. Thankfully it survived, sequestered away and preserved for us to reflect upon. This myth is significant because it represents the collective mindset of a very sophisticated civilization that evolved in America, uninfluenced by the rest of the world except through the shared objective psyche.

The *Popol Vuh,* also known by other names—*The Book of Council, The Light That Came From Across the Sea, Our Place in the Shadows*, and *The Dawn of Life*—is a metaphorical expression of the origins of the Quiché Maya culture.[254]

The Maya are an ancient group of people who have lived in southern Mexico, Guatemala, Honduras, Belize, and El Salvador from around 2000 B.C.E. to the present. There were many number of Maya tribes dotting this area, but the Quiché Maya rose to the heights of an incredible civilization in their ancient land in Preclassic time 800 B.C.E.-150 C.E. in the El Mirador Basin in central Guatemala. To help locate where this is, you may be familiar with Tikal, which is one of the cities just outside this area that was primarily a Classic city-state. The Quiché Maya were in constant conflict with the rulers of the city of Tikal, who later joined forces

---

[254] Dennis Tedlock, *The Popol Vuh,* Touchstone Book, New York, NY: 1985, 23-24.

with the fierce warriors of Teotihuacan from central Mexico to inflict harm on the Quiché.

Our focus is on the Quiché Maya, who took culture to the state level of civilization. Fifteen hundred years before the Spanish conquest, the Maya developed a sophisticated hieroglyphic system of writing capable of complex literary compositions.

In addition to their language and system of writing, the Quiché Maya developed a very sophisticated means of counting time. They conceived a calendar system that consisted of the intersection of two calendars: one was for telling the past and future, the lunar Tzolkin; and the other, the solar Haab, was used for laying out the rituals necessary to ensure proper relationship to the gods. These two calendars combined to form what is called the Long Count, allowing calculations into the far distant past and well in to the future.

The Maya were brilliant astronomers and mathematicians and developed and ability to calculate time unsurpassed by no other civilization in the world. It was they who developed the concept of zero, referring to the end or completeness rather than absence as we conceive of zero. The beginning date of their calendar and of this fourth and current world is August 11, 3114 B.C.E.

As their agrarian culture expanded to an agricultural culture, their mythology, their stories continued to reflect the interrelation-ship and dependence on the gods for healthy and productive functioning. Their gods were personifications of the seasons, weather patterns, the movement of the sun and moon across the sky. Through the movement of the celestial bodies, time was tracked, and rituals were performed to keep the cycle moving through the passage of the seasons, days, years, and aeons.

The movement of the sun across the horizon was traced from birth to death as it descended in the west at dusk. From there it went through the nine layers of the underworld, to struggle with the gods of the dead, to vanquish them and to then rise above the horizon in the eastern dawn in the form of the Maize God. Only then were the sun and moon ready for another day—a repetition of what the gods themselves had once done to start the world in motion.

Through the reenactment of the myth of creation, historical time is paused while sacred time allows it to begin anew, as Mircea Eliade so clearly describes it:

> Sacred time appears under the paradoxical aspect of a circular time, reversible and recoverable, a sort of eternal mythical present that is periodically reintegrated by means of rites. ... Sacred time does not live in the historical present. For the religious person, profane time can be periodically arrested. Cosmos is world. For the archaic cultures the world is renewed annually. The New Year comes from the sanctity of the creator's hand. The New Year is a time of purification—the expulsion of demons or scapegoats. Saturnalia is a return to chaos. ... Life cannot be repaired, it can only be recreated through symbolic repetition of the cosmogony. ... The origin myth was copied after the cosmogonic myth, for the latter is the paradigmatic model for all origins.[255]

The myths and development of culture move hand in hand, the one influencing the language and thought that is used to form the structure and guide the interactions of the people. Just as spirit and matter are inseparable, these two halves influence each other even when one side is forgotten and relegated to the shadows of unconsciousness. When this happens, the shadow takes on an uncontrollable power, forcing its recognition—usually in a destructive way.

Why did this grand, sophisticated, and flourishing New World civilization that once rivaled Egypt, Assyria, China, and the Indus River Valley suddenly collapse? By looking at the founding mythology that underpinned its culture and where it deviated from the cosmic laws and balance with *anima mundi*, we can, perhaps,

---

[255] Mircea Eliade, *The Sacred and the Profane: The Nature of Religion*, New York, NY: Harcourt, Inc. 1959, 70-84

avoid what may be a similar fate for us. We can learn from past mistakes, if we draw on the memory of past cultures.

Myth and Ritual

The Maya live according to a ritual calendar that sees time as cyclic not linear. In the ever-changing movement from one moment to another, each moment reflects the unique quality of a different god, representing a specific archetypal energy. The Maya world was created and destroyed four times, and according to their ritual calendar, it may again be destroyed and recreated as necessary in the constantly moving cycle of change. What form a new creation will take, we don't know, but we do know why the first three worlds were destroyed: a loss of relationship between the created and the divine. The gods sacrificed themselves to bring humanity into existence and in return require respect and sacrifice in their honor. This is a potent, reciprocally beneficial psychological contract that was put in place through their mythology and informed every aspect of their lives and culture.

Their myth reflects the typical union of opposites in the early matrix of chaos before energies have been differentiated. It shows how they come together again at the end of the process renewed and reunited into wholeness represented symbolically by the birth of the new god, in their case the Maize God. This new symbol for them opened the way for the creation of a cohesive culture, including spirit and soul, compassion and reason.

It is by no means an easy process because to achieve wholeness or the birth of the new god, mythic heroes had to first fight to domesticate the underworld demons, which we would say are the wild, out-of control instinctual energies that reside in the unconscious. These demon-instincts always need to be defeated before culture can emerge with its new guiding symbol, laws, and collective behaviors.

For the Maya, words connected to nature such as *sowing, dawning, sprouting* are used to refer to the dawning of this new life, light, and humanity.

186

The goal for the Maya gods is the creation of a race of beings that can contain and express consciousness and reflect it back to the gods in a more differentiated way. The gods want a race that can speak their names and worship them in the form of prayer and sacrifice:

> A fundamental aspect of indigenous highland Maya religion is the belief that human beings stand as essential mediators between this world and that of their patron deities and ancestors. Sacred ritual performed at the proper time and in a manner established by ancient precedent, is necessary to maintain this link or all creation runs the risk of collapse"[256].

The Maya words used to describe what the gods desired from humans and what was of highest value in the Maya culture included *purity of being, truth, light, clarity, whiteness,* and *brightness.* The gods wanted their creation to honor them and make offerings and sacrifices, in other words, to be in relationship with them. The creators were trying to create people who *ponder, consider*, and *have_compassion*. Those who are alive have light, while those who are dead are hidden from the sun.

Wise Council at The Dawn of Life

The *Popol Vuh* begins by saying that what will be written is the ancient Quiché word, which tells the traditions, the stories of the beginning, the recounting of their origin. The word *Quiché* means many trees, or forest. It is the account of the *sowing* and *dawning* by the creator gods at the *"Dawn of Life,"* when the world had nothing but a dark, empty sky and a calm sea below—absent was the earth. Then came the *word* that emerged from the minds of the gods.

---

[256] Allen Christenson, *Popol Vuh: The Sacred Book of the Maya*, OK: University of Oklahoma Press, Norman, 2003, 71, n. 66.

The gods came together in council and talked, thought, and pondered. Together they possessed great knowledge. Arriving at accord, they conceived of light and life, the emergence of the earth from the sea and everything that was to inhabit it. This was the word of Framer and Shaper," Heart of Sky and Heart of Earth, Sovereign and Quetzal Serpent, also called Xpiyacoc and Xmucane, He Who Has Begotten Sons and She Who Has Borne Children." Both diviners who participated in the actual creation were the divine masculine and feminine principle, "The giver of breath, and the giver of heart."[257] Together they brought the creation of the earth, mountains, valleys, trees, and waterways. Then the animals were conceived: deer, birds, pumas, jaguars, serpents, and they inhabited the land. It is said this all happened before the dawn arrived.

The Framer and Shaper asked the animals to speak, but they could not. They only "squawked, chattered, and roared."[258] The gods wanted "beings who will walk, work, and talk in an articulate and measured way, visiting shrines, giving offerings, and calling upon their makers by name, all according to the rhythms of a calendar" [259] This was a failed attempt, so the animals were left to serve and be eaten.

The disappointed gods came together again and decided this time to create beings out of earth and mud, but these merely fell apart and dissolved in water. Another mistake, but the council of gods asked the Framer and Shaper to try a third time. So next they decided to carve beings from wood, but the wooden beings did not have hearts or minds. This was not good, so the gods sent a great flood to destroy them. Along with that, the animals and instruments that had been misused by the wooden beings rose up in anger against them and crushed and tore them to pieces. (Is this a warning about our future relationship to robots?) The wooden

---

[257] Allen Christenson, *The Popol Vuh*, 2003, p. 66.
[258] Ibid, p. 76.
[259] Dennis Tedlock, *The Definitive Edition of the Mayan Book of the Sawn of Life and the Glories of Gods and Kings*. NY: Touchstone Books, New York, 1985, p. 34.

effigies that were not destroyed in the flood and the uprising were turned into monkeys.

These first abortive attempts at creation psychologically reflect what we see in adolescents as they grow into adulthood. They try on various personas as a way of seeking or approaching their true self. Many have difficulty finding it because it is not a locus; it is a process. Along the way, they first have to learn to control their animal instincts. Then, like mud that washes away, they have to contain their emotions and firm up their inner authority. Finally, the overbearing rigidity of woodenness can wound other people, just as in the myth when the household animals and instruments were furious with the wooden effigies for misusing them.

Absent was heart and related compassion. These are good lessons for the growing youngster on how to behave in the search of one's *face and heart*. This was similarly true of the growing demands of the culture, which required less selfishness and more relationship to the other, including the gods, as a collective spirit came into being and the culture matured to a new level. We parents can certainly empathize with the frustration of the gods during this part of their difficult creation process.

After the great flood at the destruction of the third world, four monstrous demons came to dwell on the earth, and the greatest of all was an arrogant and vainglorious bird called Seven Macaw. The other three were his evil wife and two monstrous sons. The text digresses from its linear movement of creation to discuss the mythic plight of the vainglorious Macaw. It was named after this beautiful and highly colorful bird set itself up in a tree as the ruler of the world, a false sun that had to be brought down. It finally was by a young set of hero twins, Hunahpú (blowgun hunter) and Xbalanque (hidden Jaguar sun), who shot out its eye of black mirror and its gold tooth, effectively blinding it and removing its power before its final fall to death.

This digression in the story offers a direct warning for the emerging Maya culture to beware of the destructive effects of arrogance and the usurpation of power. Here, it has to be quelled before creation can commence. The gods want humility, respect, and hard work. These qualities are not seen in Seven Macaw or in its two monstrous sons, who are killed next.

The sons represent opposing energies: One monster creates mountains, and the other destroys them. Without balance between these opposites there is chaos; yet out of both monsters, another myth says the earth is fashioned. The monsters represent the opposing forces that together make up the *prima material* that is required to bring forth the new ground. Out of the old and rejected comes the new creative spark, the alchemical gold, if one can remain in the psychological process long enough for the new imagination to emerge out of the struggle of opposites.

Now back to the main story: After three failed world creation attempts, the gods decided on a new tack and sent an older set of hero twins, Hun Hunaphu (one blowgun hunter) and Vacab Hunahpu (seven blowgun hunter), who were actually the father and uncle of those who defeated the Macaw, into the underworld to vanquish the demons that periodically erupted onto the earth to disrupt the balance of life. These demons were the personifications of the unconscious forces we call complexes that complicate our lives—the roaring anger, jealousy, envy, desire for attention, arrogance, and power, to name a few.

Before the final and successful creation of the world, the creation story once again revisits the effects of the wrong attitude, much like the story of the vainglorious Macaw. The world has to be cleared of all the selfish attitudes for the successful creation of the earth, establishing the proper balance between the three realms. Only then can humans be created.

Unfortunately, this older set of twins failed in their task. They did not have the proper attitude to see the tricks and shadow of the demons and were defeated by them. But the second set, their son and nephew, were actually the children of one of the demon's daughters, so they had a bit of the devil in them and could not be tricked so easily. They were more conscious of shadow and evil and were able to successfully complete their task. It is said of this creation:

> Five hundred and forty-two days later, the Maize God completed the structure of this, the Fourth Creation, by setting up the four sides and corners of the cosmos and erecting the center tree. The Maya called this tree *Wakah-Kan*, or "Raised-up-Sky."[260]

This tree is the *axis mundi* or tree of life, the center of the world around which all else circles, connecting the underworld, human realm, and spirit world all in their rightful places.

With the birth of the Maize God, agriculture took center stage, and culture moved into a solar, patriarchal phase, although the matriarchal influence remained just under the surface. Human life could be sustained in a predictable way through agriculture, and culture could develop. Yet, it was through sacrifice and blood offerings, important for ongoing life and consciousness, that the relationship between humans and the gods was maintained. "The ancient Maya offered their blood ... they returned a portion of their life force to the powers of the cosmos."[261] It is important to understand the reason for human and self-sacrifice that seems so shocking to us today. The gods created humans with their blood, and in order to be sustained, they needed humans to return it as an offering. Blood was the holy, life-force substance for both humans and gods.

As the final step, the creator gods decided to make humans in their likeness, including giving them all their own godly wisdom and vision. But, being able to see and know everything, the gods feared humans would not reproduce or work or honor them. So, the gods agreed that the mirror of humans' clear vision would be clouded with breath so they could not see everything, leaving humans to struggle to regain clarity. Even this decision, however, disturbed the gods because they were then concerned that humans would be limited to seeing only what was directly in front of their

---

[260] Schele and Mathews, 1998, p. 37.
[261] Christenson, 2003, p. 202.

faces, and they would not be able to imagine the larger reality. And how true this is!

The *Popol Vuh* itself was thought to be an instrument of clarity, and the Quiché believed they carried within their blood the potential for this divine sight. If consulted, the Book of Council would offer the individual and culture the knowledge of the cosmos and the gods. Similar themes of reducing clarity are present in Gnosticism and in the Kabbalah. In Judeo-Christian mythology the expulsion from the Garden of Eden removed humanity's clear connection to god.

In psychological terms, regaining clarity means becoming conscious of one's shadow and complexes through one's inner struggle, through the individuation process. The ego is strengthened through the collision with the outer world—through its challenges and responses, vanquishing the demons that reside in our cellars. Further, one establishes an ego-Self axis, which leads to maintaining a constant relationship to the Self, in other words, to a spiritual life that requires the correct attitude of honor and respect toward the ineffable, unknowable source of creation, destruction, and wisdom.

Yet, experience of absolute clarity returns only fleetingly when, through the experience of synchronicity, the archetypal energies break through into reality, demonstrating the unification of psyche and matter. In mythological terms this experience is a reconnection to what was lost during the creation, reestablishing the *axis mundi*.

The Rainforest Weeps

The Maya were strongly forewarned about the wrong attitude that could lead to their civilization's decline in the story of Seven Macaw, the vainglorious bird who usurped power after the destruction of the third world, and in the destructive actions of the demons in the underworld. Humbleness and respect for the sacred are the correct attitudes. Sacrifice comes from the word, *sacer*, meaning holy, or sacred. When humans fall out of relationship to the sacred, humility that comes with sacrifice is forgotten, soul is lost, and shadow escapes through an open door into our individual lives and into our cultures. We run the risk of embracing the material world as the only reality, which was Seven Macaw's and the demon's mistake. Power and striving for wealth seem too great a seduction for humans even with all the warnings, and for the Maya this took the form of frequent wars with neighboring city-states and grandiose construction projects.

There are many theories about why the Preclassic Maya civilization collapsed. It might have been due to a series of droughts, epidemics, or severe conflicts with neighboring groups, like Tikal. Richard Hansen[262], who is the lead archaeologist in charge of the excavation in the Mirador Basin where there are 26 major communities and 51 smaller cities, has theorized that as their striving for power and desire to impress themselves and surrounding cities increased, in the words of the *Popol Vuh*, their vaingloriousness increased, the Maya sought to build larger and higher temples. And in fact, the most massive pyramid in the world was built there, called *La Danta*. Each one was covered with white plaster on which they vividly painted the buildings with mostly red paint. To achieve the plaster, it was necessary to burn newly harvested green trees. Only this would achieve a heat high enough to produce the white plaster. Therefore, the Maya clear-cut a large percentage of the trees in the rainforest, destabilizing the natural balance.

The Maya at the same time built a series of raised causeways, or *sacbeob*, white roads of stone covered with lime plaster that rose above the rainforest floor. *Sac* means white, and *be* means road. These were about 6 to 12 feet high and 90 to 120 feet wide with a thick layer of plaster covering them.

This early freeway system allowed the Maya to connect to the 51 cities in the Basin as well as to transport the rich topsoil from *civales*, or swampy marshes scattered through the Mirador Basin, to replenish their agricultural fields. Yet, over about 500 years as

---

[262] Personal communication with Richard Hanson, Ph.D., in El Mirador Basin, Guatemala, November 2006.

the Maya greatly increased the quantity of lime plaster used to cover their buildings and roads, with the heavy depletion of the forests and the runoff of plaster clay into the marshes, access to water and their nutrient-rich muck used for topsoil disappeared. The *civales* were locked shut; They could no longer grow food. Conspicuous consumption caught up with them.

The rainforest is a very difficult environment to live in with rain only part of the year. While the Maya took great care in creating water systems to capture and direct the flow of water, it was the unique presence of the marshes in the Mirador Basin that supplied the precious fertile soil that enabled vast food production to support the development of the Preclassic Maya civilization. The Maya would transport the mud, the nutrient-rich material, in baskets to their cities to terrace farm vegetables, including corn, peppers, squash, and beans. The soil was so fertile they could yield two to three crops a year. It was a tenuous balance between water and population growth.

The Maya system was built on growth and expansion that funneled wealth to the nobility, ruled by the divine king. It was an inflexible system that could be thrown into imbalance easily by increased conflicts or sudden drought. With such a rigid social system, it is difficult to change course when necessary. For the Preclassic Maya, it is believed that human disruption of the balance in their natural world resulted in this environmental disaster, which led to the area's sudden collapse and abandonment in 200 C.E. The final collapse of the Maya culture in 900 C.E. coincided with the rise of the warrior state of Teotihuacan from central Mexico.

It is interesting to note that the people in the Late Preclassic period in the Mirador Basin ranged from an average height of 5 feet 8 inches to 6 feet 2 inches. The Maya stature, though, continued to decline thereafter along with their nutritious food supply.[263]

---

[263] Richard Hansen, Personal communication in November 18-21, 2013.

The *Popol Vuh* is the book that opens clear vision to the past, present, and future. The Maya daykeepers were and continue to be the shamanic priests trained to read the day and divinatory calendars, to interpret dreams, and to cure ailments. They are the ones who clear away the foggy breath on the mirror to see beyond the world of immediate reality into the further dimensions that are the domain of the gods. They remain open to the world of spirit and contemplate the natural cyclic progression of the cosmos to help fellow travelers on this life's journey relate to the multiplicity of influences at play. Parents will ask about the day of birth of a child, for example, wondering if it is auspicious or not—in other words, which god carries that particular day and what archetypal energies represented by that god will be their child's burden of fate because of that day.

We have no such book, but we do have the ability to be introspective and observant. We have our dreams that can guide our decisions and actions, if only we would pay attention to them. Through them we hear the whispering of the psyche, of the larger cosmos, offering warning and guidance, nudging us to follow our journey in balance with spirit and nature. Our inner dream voice, the Self, is our equivalent of a daykeeper. It behooves us to keep in close contact. However, when this relationship ceases to function, as it frequently does for individuals and for cultures, we lose our connection not only to the psyche but to the natural world as well. Similarly, we find ourselves in a world of materialism, power, and violence. Nature pays the heavy price for this imbalance. For the Maya, the El Mirador Basin environment became stripped and degraded to such a shortsighted extent that it could not sustain them, and this just may prove to be true for us as well.

In such times of chaos, we humans fall back on our basic survival instincts. Yet, these are shortsighted, tribal, and culturally destructive, pushing us into inflexible fundamentalist ideologies, fearful of the other. Memory is easily lost, and humanity does not learn the lessons from history as we forge ahead with our clouded decisions about the future. We, today, are in the same danger of losing ourselves as the Maya did over a thousand years ago.

But where do we *refind* or *restory* ourselves? The spirit expressed through myth, story, and ritual that originally gave shape to our culture has fallen away, leaving us without a secure sense of containment, and is now vulnerable as it crumbles into pieces around us. Materialism, superficiality, empiricism, arrogance, narcissism, loneliness are the mantras of our postmodern world where the individual is alone without cultural/ritual guidance or even the knowledge of the presence of our inner compass. Memory is forgotten as life becomes horizontal, and instead we search for utopias "out there." Many science- or science-fiction-minded folks put their hopes on artificial intelligence and technological singularity—when humans transcend biology through robots— and plan to escape to Mars when our Earth collapses, or disappear into virtual reality. We only have to rewatch the science-fiction TV series, *Battlestar Galactica,* aired in 2003, to see what this future world looks like.

Do we ever stop to ask ourselves what kind of world we want? Do we try to clear the fog on our mirror to see down the road a way? A Native American proverb about our earth reminds us: "Treat the earth well. It was not given to you by your parents, it was loaned to you by your children." How do we reconcile this reminder with our increasing reliance only on science and technology that got us into this mess to begin with?

There is a striking similarity between the lead-up to the Maya collapse in 200 B.C.E. and what we are facing today. While their impact was limited to the size equivalent to one country and one rainforest, our current risk of collapse will affect the entire world and all living creatures, the very soul of the world, *anima mundi.* We are not merely dealing with the effects of plaster runoff but instead runoff of chemicals and carbon that have grown far beyond our control, affecting every living organism on our precious earth.

Conclusion

The Maya gods came together in council to make decisions on the creation of the world. This is a good reminder to share the decision-making process with others—in honest relationship, listening to what others have to say with tolerance and under-standing. Yet, it not only refers to outer real others but also to all the inner others in the unconscious whose voices influence the decisions we make on a daily basis. Sitting in council with all of them would help bring the multiplicity of viewpoints, complexes, desires, and affects into view to enable us to make the best possible decisions, even if it means we have a few abortive creations. We learn from those, too.

Myths speak with the loud as well as the soft voice. In the *Popol Vuh*, it states that at the successful creation of humans and this world the Maya gods were turned to stone when the sun's first light touched the face of the earth. Their voices became internal so that humans might remember their origin and connection to the living gods through their dreams and that soft inner voice of wisdom that whispers when they dare to listen.

The Maya see their creation myth as laid out in their Book of Council as the road back to their gods. Understanding it opens the door to the clarity of vision dimmed at the creation of humans. From another culture, another time, we can strive for that same clarity by understanding the ancient voices of wisdom—from our own ancestors, as well as from those of other cultures. The one voice speaks to all through metaphor from the archive of humanity's memories. By listening and reflecting, we can return to the well at the dawn of life for renewal.

CHAPTER

18

From Transiency to Healing in the Postmodern Condition

Jung's *Red Book*, which took nearly a century to let its images emerge from incubation in the cave of treasures, is a story of healing, of Restorying the Self, of which we shall now hear more.

The Red Book and the Spirit of the Depths
Nancy Swift Furlotti

How quickly the Western world has changed since the 1800s, the height of the Industrial Revolution, when our grandparents and Jung moved into the modern era and technology was born. Unbeknownst to anyone at the time, it would take on a life of its own. Two hundred years later, our world is ruled by cell phones, computers, social websites, like Facebook and Twitter, sound bites that move faster and faster, and a wealth of information that is available in the remotest parts of the globe, like a flow of water that fills every empty crevice. We watch old movies and comment on how slow they are. We move through our lives and the world virtually connected, but in reality, alone, in our cars, or cubicles, taking online classes, ordering groceries to be delivered, watching movies alone on Netflix, texting, checking emails 24 hours a day. Always connected, but less and less so in reality.

Teenagers will sit together in restaurants and text each other rather than communicate directly. They use texting as an immediate release of emotion and expect it in return. Everything is immediate gratification, wanting to be in touch NOW[264]. The bings, bells, or buzzes that announce a text or email stimulate the dopamine receptors in the brain much like cocaine or other drugs, establishing addictive behavior. We long for the next "fix." The expected immediacy impinges on our lives, exacerbating our obsessions and compulsions, taking us away from our quiet, thoughtful, introspective moments, the reverie of intimate communication.

It has recently been revealed how the tobacco and food industries actively seek to make their products more addictive, how advertising focuses on stimulating the brain to encourage us to buy certain products. We can now add technology to that list. We are addicted. Imagine what will happen when robots are designed to self-replicate, and the superdromes decide we humans are the enemy! A mathematician from MIT, Norbert Weiner, offered a warning back in 1949:

> The machines will do what we ask them to do, not what we ought to ask them to do. ... There is a general agreement among the sages of the peoples of the past ages, that if we are granted power commensurate with our will, we are more likely to use it wrongly than to use it rightly, to use it stupidly than to use it intelligently. ... Moreover, if we move in the direction of making machines which learn and whose behavior is modified by experience, we must face the fact that every degree of independence we give the machine is a degree of possible defiance of our wishes. [265]

---

[264] Sherry Turkle, *Alone Together: Why We expect More From Technology and Less From Each Other,* Basic Books, New York, NY: 2011, 174.

[265] J. Markoff, In 1949, He imagined and age of robots (The New York Times, May 21, 2013).

That is pretty darn scary but true. Dmitry Itskov, as of this writing, a 32-year-old Russian multimillionaire, online media magnate, has a colossal dream, to upload the human brain's contents to cyborgs he calls avatars. According to recent a New York Times article about him:

> Scientists are taking tiny steps toward melding humans and machines all the time. Ray Kurzweil, the futurist and now Google's director of engineering, argues that technology is advancing exponentially and that human life will be irreversibly transformed to the point that there will be no difference between human and machine or between physical and virtual reality.[266]

A number of years ago I attended the Aspen Brain Forum, which focused on early learning. One of the speakers stated that American children were the second-most unhappy in the developed world. Child psychologist and researcher/author Susan Engle emphatically reminds us that we are social creatures, that babies need a real person to interact with, not a TV or touch screen, and that there may be a critical period within which children learn empathy. And if you put two babies together, they learn twice as fast. Technology can steal our moments of creativity and play. On the other hand, it is reported that certain video games can stimulate the brain to increase attention, and a study currently underway at UCLA is testing the use of games in reducing the symptoms of Obsessive Compulsive Disorder. Technology may increase this behavior as well as providing its cure. This is quite alchemical. There are definitely positives for learning if the right technology is used at the right time. There has to be a balance between the opposites, though, which is what Jung stressed, pointing out that "man has achieved a wealth of useful gadgets, but, to offset that, he has torn

---

[266] D. Segal, This man is not a cyborg. Yet (*The New York Times*, June 2, 2013). https://www.nytimes.com/2013/06/02/business/dmitry-itskov-and-the-avatar-quest. html

open the abyss, and what will become of him now—where can he make a halt?"[267] There are those so caught up with the cyber archetype that they are driven on.

A 2010 quote in *Wired* magazine caught my attention, "Technology is something that can give meaning to our lives, particularly in a secular world"[268]. Kevin Kelly started *Wired* and goes even further in his book *What Technology Wants*[269] when he describes technology as the *technium*, giving it a name that carries movement as a system:

> The technium is less an adversary to life than its extension. Humans are not the culmination of this trajectory but an intermediary, smack in the middle between the born and the made. ... The technium is the way the universe has engineered its own self-awareness. ... Someday we may believe the most convivial technology we can make is not a testament to human ingenuity but a testimony of the holy. As the technium's autonomy rises, we have less influence over the made. It follows its own momentum begun at the big bang. In a new axial age, it is possible the greatest technological works will be considered a portrait of God rather than of us.

Now, who is projecting the Self onto technology in this secular age? This is really quite unbelievable and, of course, extreme, or is it? The danger is that we move closer to the world of Cylons, those ever-evolving robots in the Science Fiction TV series *Battlestar Galactica*, that began to develop emotions and threatened to evolve beyond humankind. Even the Cylons respected the stories of the past and developed more of a spiritual orientation than our technological secularists of today.

---

[267] CW 9i par 454.
[268] (Wired Magazine, October 2010) 124.
[269] Kevin Kelly, *What Technology Wants*, Viking, Penguin Group, New York, NY: 356-358.

Karen Armstrong, in her book *The Case for God*[270] points out that:

> The Christian thinkers of the past understood faith primarily as a practice, rather than as a system—not as something that people thought but something they did. Their God was not a being to be defined or a proposition to be tested, but an ultimate reality to be approached through myth, ritual and "apophatic" theology ... emphasizing that we cannot know the divine. The Christian West has largely lost, and the rise of modern science is to blame. Not because science and religion are unalterably opposed, but because religious thinkers succumbed to a fatal case of science envy.[271]

Recently, the rewriters of the Diagnostic Statistical Manual, used in the United States to diagnose psychological symptoms and illnesses, actually wanted to include introversion as a pathology in the new edition! Thankfully that was stopped by a group of outspoken Jungians. Even Pico Iyer, the soulful travel writer, recently wrote an article titled, The Joy of Quiet, which espouses the importance of introversion and the need to escape the constant stream of information. The negative side of our scientific technological advances moving at the speed of light take us away from what is precious to being human, the soulful recognition of our individual myth of who we are, our story, and how we are connected to the divine.

In *The Red Book*, Jung demonstrates, through his own inner journey, his discovery of the way back to soul, through a quieter, slower, introverted exploration of the inner world and its connection to the whole. It was not painless or pleasant by any

---

[270] Karen Armstrong, *A Case for God*, Anchor Books, Random House, New York, NY: 2010.
[271] R. Douthat, Perpetual Revelations: How Ideas and Practices from the Past Might Guide Believers in a Modern Approach to God. *New York Times Book Review*, October 4, 2009.

means, quite the contrary. It was very personal as well as collective, and vital, in the true sense of the word, to his life. He spent the remainder of his life and work making sense of it for himself and for us. I quote from *Liber Novus, The Red Book:*

> Your soul is in great need, because drought weighs on its world. If you look outside yourselves, you see the far-off forest and mountains, and above them your vision climbs to the realms of the stars. And if you look into yourselves, you will see on the other hand the nearby as far-off and infinite, since the world of the inner is as infinite as the world of the outer. Just as you become a part of the manifold essence of the world through your bodies, so you become a part of the manifold essence of the inner world through your soul. This inner world is truly infinite, in no way poorer than the outer one. Man lives in two worlds.[272]

As we move back into the world of our outer reality today, I am curious about the upsurge in the prevalence of autism. The last generation was sped up by ADHD, and in this one, one in 100 children have some form of autism. Evidently, there is a genetic component to autism, mutations in the genes, but no one is saying why the genes are mutating. Through genetics we are learning that our genes continue to mutate and are mutating at a faster pace than ever before. Is there a connection between Asperger's, a mild form of autism, now called Autism Spectrum Disorder, and IT, meaning technology? This question is taken up by the magazine, *Computerworld:*

> We wouldn't even have any computers if we didn't have Asperger's. ... All these labels—"geek" and "nerd" and "mild Asperger's"—are all getting at the same thing. ...

---

[272] C.G. Jung, *The Red Book*, Norton & Company, New York, NY: 2009, 264.

The Asperger's brain is interested in things rather than people, and people who are interested in things have given us the computer you're working on right now.[273]

Perhaps Jung would ask which came first, the speeding-up of technology or autism? What is the shadow of Asperger's with its focus on technology and objects rather than people? Would it be soulful interactions that bring the warmth of touch; would it be clear communication between two people sitting face to face, not on Facebook; would it be recognizing the need for attachment and the slow pace it takes to achieve that; would it be introversion and self-exploration? Children need these things for healthy attachment and development. And so do adolescents, 45 percent of whom now self-describe as lonely, while 50 percent say they are addicted to their cell phones. They watch others' perfect lives on Facebook, while falling into the depression of their own mundane reality. Doctors don't even touch anymore, hiding behind a computer screen.

It is this trend of moving further and further away from the actual other and ourselves that Jung abhorred, that he struggled to expose for the health of individuals and cultures, as well. He framed it in terms of an uprootedness that has given rise to the discontents of civilization. We rush ahead looking for a golden age, running from an increasing sense of insufficiency and dissatisfaction. We have lost our means of imagining our story or connecting to the stories of the past. Mythology as collective stories provides a synthesis of belief with practical knowledge. It roots us in the past through its narrative and encourages the continuing imagination of the story of our creation, creating onward. Remembering the past and the stories, we at least are reminded to ask the fundamental questions. What are we creating, and is this really what we want to create for our human race?

---

[273] T. Mayor, Asperger's and IT: Dark Secret or Open Secret? *Computerworld*, April 2, 2008. http://www.computerworld.com/s/article/9072119/Asperger_s_and_IT_Dark_secret_or_open_secret_?taxonomyID=14&pageNumber=2

In Jung's world of 1913, when he began *The Red Book*, Europe was quickly descending into chaos and destruction, evil forces were released from the unconscious, preparing to perpetrate unspeakable crimes. He fell into a depression that forced him to descend into himself to explore the un-chartered territory of the psyche. Jung had lived his life fully out of personality number One. He was a successful and famous psychiatrist, married with five children, had a full analytical practice, and yet this was not enough. He was beckoned inward in search of his personality number Two, whom he had left behind years before:

> My soul, where are you? Do you hear me? I speak, I call you—are you there? I have returned, I am here again. I have shaken the dust of all the lands from my feet, and I have come to you. I am with you. After long years of long wandering, I have come to you again. Should I tell you everything I have seen, experienced, and drunk in? ... I will wander with you and ascend to my solitude.[274]

Through his depression, Jung struggled to pull away from what he called *The Spirit of the Times, which* represented his extraverted, empirical, scientific attitude, presented in his successful, hardworking persona of personality number One. Instead, he followed the softer beckoning voice of *The Spirit of the Depths.* It was this voice that led him back to his lost and forgotten soul. Although not an easy task to reconnect with her, he did, but not before journeying into Hell which means becoming Hell himself; falling into a Divine Madness, murdering his heroic side so that he could be open to receive the gift of the creative, new life. As Jung describes it:

> Hell is when you know that everything serious that you have planned with yourself is also laughable, that everything fine is also brutal, that everything good is also bad, that everything high is also low, and that

---

[274] C.G. Jung, *The Red Book*, Norton & Company, New York, NY: 2009. 232

everything pleasant is also shameful. ... But the deepest Hell is when you realize that Hell is also no Hell, but a cheerful Heaven, not a Heaven in itself, but in this respect a Heaven, and in that respect a Hell.[275]

For Jung, the ambiguity of God was revealed within him, the ambiguity of life itself. This ambiguity that is in the chaos of our inner worlds is what we turn away from, what we run from, why we turn our faces toward the outer world, toward others and things. Jung says: " May the frightfulness become so great that it can turn men's eyes inward, so that they will no longer seeks the self in others, but in themselves ... fundamentally you are so terrified of yourself ... you prefer to run to all others rather than to yourself."[276] Shalit[277] reminds us that along with this movement outward, there is a speeding-up of extraverted activity resulting in a "fleeing of the center" leaving no room for depression or deep impressions.

Can we listen to what is trying to be born within us? Torn between the chaos and the newly emerging symbol? Depression is lifted with medications, which are now seen as mere placebos, and one never lands long enough to form an impression. As one becomes remote from the reality of oneself, one loses connection to the emotions that keep us connected to or concerned with others. We live behind a screen, remain untouched, and do not touch. Without the struggle with the very real other, a solid ego is never formed, but instead remains fragile and vulnerable to the biding of unconscious forces within, including what Jung termed our *shadow* and *complexes*. This is our chaos. These are the aspects of our unlived lives, those parts of ourselves that have not been brought into consciousness to strengthen and fill out our ego, those elements of ourselves that are required for us to flower into the

---

[275] ibid 244
[276] Ibis 254
[277] Erel Shalit, 2010 p. 5

unique personality we are intended to be. We remain transient, not here or there.

In *The Red Book*, it is precisely these aspects of himself Jung faced in his conversations with his myriad inner figures. This was not easy going for him, nor is it for anyone who undertakes this journey. It includes sacrifice, humiliation, deflation, not knowing, loss, sadness, despair, and meeting your own devil—that is your shadow side. Through the process, parts of oneself are changed: Some recede while others emerge in prominence. If taken seriously, there is a movement toward greater balance, less neurosis. Jung says: "It is the self that causes me to make the sacrifice; nay more, it compels me to make it. The self is the sacrifice, and I am the sacrificed gift, the human sacrifice."[278] "Sacrifice always means the renunciation of a valuable part of oneself, and through it the sacrificer escapes being devoured. In other words, there is no transformation into the opposite, but rather equilibrium and union, from which arises a new form of libido."[279] Here, Jung is referring to the transcendent function, the ability in each one of us to bring together the opposites, holding the conflict that is created, standing with our feet in the fire long enough for something new to emerge. It is struggling with one's shadow, with one's own personal devil.

Jung took his inner figures very seriously. At times he tried to run from them, to deny them, and push them away, but in the end, he had the fortitude to stand his own ground and face them. They were real because they affected him and his world. He was touched by them, and he touched back. Jung states: "I am of the opinion that the psyche is the most tremendous fact of human life."[280] Elijah tells Jung in *The Red Book*, "We are real, and not symbols."[281] This is a very powerful and very important statement. Later, Jung goes further to say:

---

[278] C.G. Jung CW II para. 397.
[279] C.G. Jung, CW6 para. 339.
[280] C.G. Jung, 9i para. 206.
[281] C.G. Jung, *The Red Book*, p. 246.

Fantasy is reality. ... Images are the only reality we apprehend directly; they are the primary expression of mind and of its energy which we cannot know except through the images it presents. When we perceive a fantasy image we are looking in to the mind of instinct, seeing the libido itself... Fantasy as imaginative activity is identical with the flow of psychic energy.[282]

Jung, through arduous inner work, discovered an antidote to the postmodern malaise now threatening to breach the gates into ever-evolving technological nightmares that may leave us all in a state of transiency, at best. It is not an easy remedy but one that results in the transformation of the personality, the coming home to the soul of one's being. One experiences solid ground rather than the illusionary quicksand that is the ground on which our postmodern, technium culture rests. Jung brings the multiplicity within back to a unity, which he calls the Self. This process of healing is individual and cultural, as well. We are each one of the multiplicity of the larger whole. With each person's change of attitude and greater consciousness of the monster, we may be bringing into existence, by default, a larger cultural change. If the alchemical dictum is true, that the inner mirrors the outer, the microcosm mirrors the macrocosm, then we all need to accept responsibility for our future.

From Jung's early prewar visions of the frozen land where nothing grows and all is stagnant, he saw a *Tree of Life* emerge. Jung's vision was a foreshadowing of what was to become of Europe in 1913 but also pointed to Jung's own need to break through the frozen aspects of himself and descend into the center of his being to understand and assimilate his own shadow contents in the unconscious. While he was a pioneer in exploring his inner "cosmos," his example is a continuing gift for the world, as well. A new *Tree of Life* could grow. Through his journey inward Jung,

---

[282] C.G. Jung, CW 11, para: 769, 889., CW 8 para. 389. CW 6 para. 722, 711.

realized that his process mirrored the template for what he later called *individuation*, or the natural course of human development. He challenges each one of us to embark on our own unique journey to find ourselves and not remain satisfied living the transient life with no say about how we or the world continues to be created. I will end with Jung's own words:

> Not every opinion of any individual contemporary will decide the truth and error of what has been discovered, but rather [it will be decided by] future generations and destiny. There are things that are not yet true today, perhaps we are not yet permitted to recognize them as true, although they may be true tomorrow. Therefore every pioneer must take his own path, alone but hopeful, with the open eyes of one who is conscious of its solitude and of the perils of its dim precipices. Our age is seeking a new spring of life. I found one and drank of it and the water tasted good. That is all that I can or want to say. My intention and my duty to society is fulfilled when I have described, as well as I can, the way that led me to the spring.[283]

I hope we can each find our own spring and collectively question, challenge, and have opinions about the story we remember and want to tell of our world.

---

[283] C.G. Jung, Collected Papers on Analytical Psychology, (New York, NY: Moffat Yard and Co. 1917). pp. 443-4.

CHAPTER

19

Will Fishes Fly in Aquarius –
Or Will They Drown in the Bucket?

Aquarius, the 11th sign of the Zodiac, is symbolized by the water-carrier. While Pisces is a water-sign, Aquarius is an air-sign, associated with ideas and intellectual pursuit rather than physical matters; intuition and thinking rather than sensation and feeling. Therefore, the transition from Pisces to Aquarius is one from water to air, as reflected in the dream of one woman, who in the dream has:

> Just given birth in a natural pool at the slopes of a mountain. She picks up and holds the newborn baby in her arms, walking a path up around the mountain. Suddenly the infant points his little finger and looks toward the sky, saying, "look, fishes fly in the sky![284]

Man is no longer the fish in divine waters or the fisherman at the daybreak of history. Man now holds the bucket, the containing vessel into which the matter and the image of the universe have been amassed. In a way, the Self and the God-image are in the hands of man's ego and consciousness. This creates

---

[284] A female patient.

opportunities as well as dangers, such as cloning and computers made of living organisms. With cloning, for instance, the genetic blueprint of the cloned individual is selected and determined by the human craftsman—probably the very craftsman God had warned against. In *The Ethics of Human Cloning*, Leon Kass declares how we are turning man into:

> Simply another one of the man-made things. Human nature becomes merely the last part of nature to succumb to the technological project, which turns all of nature into raw material at human disposal. ... Such an arrangement is profoundly dehumanizing, no matter how good the product. Mass-scale cloning of the same individual makes the point vividly; but the violation of human equality, freedom, and dignity is present even in a single planned clone.[285]

In his article "Why the Future Doesn't Need Us," Bill Joy, cofounder and chief scientist of Sun Microsystems, says that he feels "very uncomfortable" about his enabling "the construction of the technology that may replace our species." And he goes on to say, "We have yet to come to terms with the fact that the most compelling twenty-first century technologies—robotics, genetic engineering, and nanotechnology—pose a different threat than the technologies that have come before. ... They can self-replicate."[286] That is, the buck does not stop, but what man-the-God has created can now self-replicate. As an ominous and gloomy example, this reflecting scientist warns against constructed "bacteria" getting out of control, "out-compet[ing] real bacteria: They could spread like blowing pollen, replicate swiftly, and reduce the biosphere to dust in a matter of days."[287]

---

[285] Leon Kass & James Wilson, *The Ethics of Human Cloning*, AEI Press, Washington, DC: 1998, 38-39.

[286] Bill Joy, Why the Future Doesn't Need Us, *Wired*, 2000, wired.com/wired/archive/8.04/joy.

[287] Bill Joy.

At the beginning of the 20th century, Freud told us, by means of his story of hysteria, that behind the girdle there is libido.[288] Later, Kafka put his magnifying glass on the meaninglessness and alienation that took the place of repression.[289] Having crossed the threshold of the 21st century, we need perhaps to imagine the *nothingness* beyond man.

Matter is becoming minimized *ad infinitum*. Consider, for instance, the space that is needed to gather all the undifferentiated material scattered and brought together on the internet—text, pictures, pornography, theology, garbage, and treasures. In 1999, all it would take to store everything on the net on compact discs amounted to about 4 cubic meters. According to one calculation, in 2012, the hundreds of millions of hard drives that store information on the internet would together amount to less than the capacity of one oil tanker.[290] So even considering the considerable growth, the relationship between quantity of information and physical space has altered drastically.

In *Memories, Dreams, Reflections* Jung writes:

> ... at the end of the second millennium the outlines of a universal catastrophe became apparent, at first in the form of a threat to consciousness. This threat consists in giantism—in other words, a hubris of consciousness —in the assertion: "Nothing is greater than man and his deeds."[291]

Truth is, that *nothing* **is** greater than (hu)man.

---

[288] Josef Breuer and Sigmund Freud, *Studies on Hysteria*, Penguin, Harmondsworth, England, 1991; Shalit, *The Complex*, 7.

[289] E.g. The Metamorphosis, in Franz Kafka, *The Complete Stories*, Schocken, New York, NY: 1971, 89-139; Franz Kafka's Letter to Father, in Shalit, *The Complex*, 92-103.

[290] Randall Munroe, http://what-if.xkcd.com/23/. However, we need to consider that growth is exponential; even if less will store more, the physical space the internet requires will grow, likely till a point when it will be replaced by something entirely different.

[291] C.G. Jung, *Memories, Dreams, Reflections*, Vintage, New York, NY: 1965, 328.

In his *Seven Sermons to the Dead*, Jung says, "Nothingness is the same as fullness. In infinity full is no better than empty. Nothingness is both empty and full. This nothingness or fullness ... is nothing and everything ... both beginning and end of created beings."[292] Later on, Jung makes it clear that "God and devil are the first manifestations of nothingness,"[293] and in fact, just as empty is no different than full, God and devil are one, until consciousness sets them apart. The Hebrew word for God, *El* (אל), means both strength, and—*nothing*. While "for the gods who know no death, nothing is serious,"[294] for us human mortals, nothing *is* and needs to be serious.

If man is convinced that *he*, or *she*, the human being, is greater than nothing, then the *no*-thing, the man-made golem in whatever shape, e.g., organic computer or cloned creature or atomic leaks in the water-carrier's bucket, will rise up against man and destroy him. "The forces that hold the fabric of the world together have got into the hands of man, so that he even has the idea of making an artificial sun," says Jung, commenting on the atomic explosion. And, he continues, "God's powers have passed into our hands, our fallible human hands. The consequences are inconceivable. The powers themselves are not evil, but in the hands of man they are an appalling danger—in evil hands."[295]

When Godlike powers pass into human hands, hubris may destroy human values. Nazism was an evil "cure" that turned people trained to be healers into executioners, using medical deception and distortion, by means of which man "journey[d] toward nothingness," as Primo Levi calls the transportation by

---

[292] Jung, *Memories, Dreams, Reflections*, 379. Howard Pollack has made me aware of the similarity between this passage from the *Seven Sermons to the Dead* and the *Heart Sutra* (The Heart Sutra Home Page, http://members.ozemail.com.au/~mooncharts/heartsutra/english.html).

[293] Jung, *Memories, Dreams, Reflections*, 382.

[294] *Dictionary of the History of Ideas*; http://etext.lib.virginia.edu/cgi-local/DHI/dhiana.cgi?id=dv3-71.

[295] Jung, CW 10, ¶879.

train toward extermination.[296] Some years ago, I went to Treblinka, where more than 800,000 Jews were murdered in the span of 13 months. The "health-clinic" to which the elderly, the sick, and children who could not run quickly enough up the *Himmelstrasse*, the pine branch-decorated, deadly so-called "Road to Heaven" (i.e., the shortcut to the gas chambers), this *Lazarett* to which they were brought was a mere deceptive facade, behind which they were shot and pushed into the constantly burning pit of corpses. And the clock at the as-if railway station, which, as Hannah Arendt says, "looked exactly like an ordinary station anywhere in Germany ...; it was a perfect imitation,"[297] the clock always showed 6, no matter what time of the day you arrived for extermination.[298] By deception, the time of humanity had come to a standstill as man's evil mind turned man into dust around the clock.

Will Aquarius the water-carrier be a sorcerer who creates illusory images behind which hide loneliness, meaninglessness, and nothingness, and sometimes evil deception? Or will we be able to hold the nothing, the *no*-thing, having the wisdom to realize our decisive role in shaping the universe of the future and the enormous responsibility it puts on our shoulders? As the Maggid of Mezritch is quoted to have said:

> Nothing in the world can change from one reality into another, unless it first turns into nothing, that is, into the reality of the between-stage. In that stage it is nothing and no one can grasp it, for it has reached the rung of nothingness, just as before creation. ... And philosophy terms this the primal state which no one can grasp because it is a force which precedes creation; it is called chaos. It is the same with the sprouting seed. It does not begin to sprout until the seed disintegrates

---

[296] Primo Levi, *If This Is A Man* (London, England: Abacus, 1987), 23.
[297] Arendt, 89.
[298] Cf. Yitzhak Arad, *Belzec, Sobibor, Reinhard, Treblinka: The Operation Death Camps* (Bloomington IN, Indiana University Press, 1999), 122.

in the earth and the quality of seed-dom is destroyed in order that it may attain to nothingness which is the rung before creation. And this rung is called wisdom, that is to say, a thought which cannot be made manifest. Then this thought gives rise to creation, as it is written: "In wisdom hast Though made them all."[299]

The no-thing would then be based on the invisible connection of individual souls respecting the *anima mundi*, the World Soul, and the *magna mater*, Great Mother, or the *Great Matter*. Or, as expressed in one woman's dream, on a theme recurrent among several of my analysands, in which she tells:

I look into the air, where I see an endless flock of migrating birds. I am amazed at the remarkable order of their flight and coordination between them, forming one unitary whole. I realize that the source of the wonder is the depth of their connection.

The soul has been extracted from the waters of below ("water" in Hebrew is *maim),* to fly in the waters of above ("heaven" in Hebrew is *shamaim*). If the soul shall not drown in the bucket, "God's powers having passed into our hands," as Jung said,[300] man must establish a soulful connection—between human and human, man and woman, human and nature, and between the time of the human ego, *Chronos*, and divine time, *Aion*. This, I think, would be a connection between *anima mundi*, the world psyche, and *eco mundi*, the world home, amounting to a psycho-ecology.

I imagine this can be done by means of consciousness and respect. To respect is to re-spect, to look again, i.e., looking another time, looking carefully, consciously, critically, and respectfully,

---

[299] Martin Buber, *Tales of the Hasidim*, Book One, Schocken Books, New York, NY: 1991, 104.
[300] Jung, CW 10, ¶879.

again and again—that is, *to re-spect*. To respect and to relate is the essence of soul. As Jung writes:

> Out of the decay the soul mounts up to heaven. Only *one* soul departs from the two, for the two have indeed become one. This brings out the nature of the soul as a *vinculum* or *ligamentum*: it is a function of relationship.[301]

The hope for the future, then, is our capacity to protect the soul that mounts up to heaven, out of the decay, to keep a watchman's eye on the flow of images that ascend from the depreciated matter.

---

[301] Jung, CW 16, ¶475.

CHAPTER

The Council of Elders

The face of democracy is changing; democracy is not only, or not even primarily, elections, but a state of mind. While with the advent of the internet and social networks, blogs and talkbacks, the opportunity to express oneself in whatever fashion—including fake news, lies and deceit, and however vulgar the manner—is no longer the exclusive domain of manipulative politicians, but belongs to everyone. This may seem democratic, but it does not pertain to a democratic state of mind, which implies "life, liberty and the pursuit of happiness." The word democracy comes from the Greek word, *demos*, which means people, and includes fairness for all, not just the few. It is government run by the people, for the people. It is not just a government structure or the ability to vote, but is, in essence, the way we approach our daily lives with respect for others.

What we see today is a loss of democracy and a world that is rather anarchistic, meaning that the ability to express oneself is being coopted as words are being twisted and presented in manipulated, self-serving ways, creating fake news to influence voters toward one political party or another, toward one viewpoint or another. There is no longer a mutual respect for opinions or opposing sides in dialogue.

Furthermore, with the emergence of global megacompanies in the virtual world, which by means of their capability to search and gather and connect friends and strangers in the "globosphere," they have become gatherers of information about their customers, and that is us. "It is reported that you (Facebook) can collect information on the feelings and emotions of your members and then target them—in the never-ending quest for another dollar of revenue."[302] It seems that devices like Amazon's Alexa can and do record conversations and send them to other people. Big Brother is no longer the only totalitarian state, but Google, Facebook, and Amazon are now players in that role by having established monopolies on information and how it is disseminated. Our lives are being monitored and shaped not according to free will but are being coopted for the benefit of profits and shareholder values.

Companies are allowing AI to speak for them, and we humans cannot tell the difference between a bot and a person. As a 2017 Harvard study warned, "The existence of widespread AI forgery capabilities will erode social trust, as previously reliable evidence becomes highly uncertain."[303] Free speech is an essential component of democracy. When that is diluted, freedom is lost along with the essence of democracy

Since the turn of our century there has been an exponential growth in technology, leaving humans across the globe in the dust, blinded by what is occurring, choking on what is left behind, moving so incredibly fast, following Moore's law, faster than the human mind can adapt. What we see as the dust clears is a changed world, one that is no longer recognizable in so many ways. We know that change frequently causes fear, as a basic human instinct, and fear causes us to grab what looks like a lifeline, but that lifeline turns out to be the very thing that takes away our freedom,

---

[302] MeWe ad, Dear Facebook. In *The New York Times*, National, Sunday, May 27, 2018, p. 19.

[303] Bill LaPlante and Katharyn White, Five Myths about Artificial Intelligence (In *The Washington Post*, Outlook, April 27, 2018). https://www.washingtonpost.com/outlook/five-myths/five-myths-about-artificial-intelligence/2018/04/27/76c35408-4959-11e8-827e-190efaf1f1ee_story.html

democracy, and compassion for others. We have seen it before in the form of fundamentalism/fascism, and we are there once again. Book agent Erik Hane who writes at the top of his voice, continues:

> It is one of fascism's goals to monopolize our attention. It would like to shrink our imagination. ... Fascism welcomes or attempts to play logical "gotcha" with it inconsistencies because it knows we will lose—not because we won't find a fallacy but because the fallacy won't matter.[304]

How easy it is to give up our rights, our voices, and sink into numbness, navigating our postmodern transient world as superficially as we can, not to have to feel what is real. In this transient space we don't hold onto ourselves in any substantial way, having lost our connection to soul, to the Self, to our imagination. We can't imagine what will be coming next, except the prefab fantasy of more reality TV.

Is there any guidance for living in our brave new world? In the Maya *Popol Vuh*, it was the council of gods that led to the creation of this world. They made mistakes along the way—three worlds were failures—but they finally got it right. The collaboration, mutual respect, and perseverance led to a successful result, a result that was built on mutual respect between the gods themselves and between the gods and humans. It was a democratic process. Where is our council of elders to guide and lead humanity now in an ethical, moral, mutually respectful way to a more sustainable and constructive future? We need these elders as role-models and as guides in all areas of culture and life to ensure a healthy, productive future, but alas they are sorely missing. The elders in the Maya myth made humans to be like them in all ways. The wise elder resides within each one of us, if only we take the time to clear away the breath on the mirror, to find him or her to engage in dialogue.

---

[304] David Brooks, Trump's Magical Fantasy World. In *The New York Times* op-ed, Friday May 25, 2018, p. A21.

Think Tanks and Forums of Ethics

Just as medical experiments require approval from the U.S. Food and Drug Administration and the Helsinki Committee, similar oversight should apply to technological inventions and developments. We live in a brave new world where, according to Elon Musk, "Artificial Intelligence is more dangerous than nukes," and Stephen Hawking expressed similar fears of robots replacing humans. With technological advances happening at increasing speeds, are we giving attention to the potential negative and even dangerous outcomes, the shadow side of all this rapid development as we race to create the next golem? Have we given thought to how to avoid programming our shadow into AI—our envy, fear, insecurity, paranoia, aggression—the destructive side of our humanity. As a culture, we certainly have not figured out how **not** to project these inferiorities onto others while they remains the unseen, insidious part of humanity that cultures and religions throughout the world have tried to tame, unsuccessfully. While AI may not actually be the initial danger, it is hard to imagine robots will be docile, empathic entities, especially after they learn to think on their own and realize the true nature of the other race they are dealing with—humans.

The big question remains: Who is responsible for ensuring that technological developments are not dangerous? Journalist John Thornhill returns to philosophy to try to make sense of this question:

> Such questions have previously been the realm of moral philosophers, such as Professor Michael Sandel at Harvard, when discussing 'is murder ever justified?' But now boardrooms and car owners—may increasingly find themselves having to debate the merits of Immanuel Kant's categorical imperative versus the utilitarianism of Jeremy Bentham.[305]

---

[305] John Thornhill, Brave New Technology Needs New Ethics, (In *Financial Times, Technology*, January 20, 1916). https://www.ft.com/content/dd328bf4-a25e-11e5-8d70-42b68cfae6e4

Kant ethical standpoint focused on the duty of the individual to do what was considered universally right in terms of actions as opposed to consequences. In contrast, Bentham believed in a utilitarianism approach to ethics, which focused not on rules but on the consequences rather actions.

There are a number of important ethical implications that need serious consideration including: 1. how technology is developed; 2. the relationship between technology and our human values; and 3. how it affects our well-being as humans and our ability to flourish on our planet, Earth. As we struggle with the questions and implications of our swiftly evolving postmodern reality, it sadly seems the safety of our future is being left up to the boards of tech companies and futuristic sci-fi thinkers interested in planning utopias rather than defending our precious planet and all its many inhabitants, including the animals and plants.

Nevertheless, there are individuals trying to increase awareness of the dangers of technology. For example, some teachers in grammar schools discuss personal, security issues with their young students. Yet, our politicians seem too deeply embroiled in their own power grabs to focus on these important issues facing the future of humanity. We can only hope that the awareness for the need to have think tanks and ethics committees focusing on these issues will increase collectively so that humanity will be able to enjoy the benefits of AI without all the potential dangers. The opposites in this potentially perilous trajectory need to find a balance through conscious consideration.

Conclusion

In the recollectivized world and psyche, the ego becomes associatively adaptive, engulfed by memoryless transiency, driven by the need for instant gratification and impatient, restless movement. This is because the ego is not anchored in an archetypal sense of meaning, anchored in the archetype of meaning and wholeness, the Self. It is by self-recollection, as Jung says, by gathering together what has been scattered, that we become human.

Insofar as life moves along an axis from *experience* to *internalization,* in other words, *consciousness* and *recollection* of the experienced life, the older person integrates his or her life by self-recollection. This process is a continually repeated cycle, day and night. The Transient Personality, on the other hand, who thrives so seemingly well in our fragmented postmodern condition, will remain bleeding by the scars of those shattered fragments that he or she hardly felt when stepping on them, just like the post traumatic sufferer who is haunted by the sometimes unintegratable horrors that shattered the world and his or her psyche.

Self-recollection pertains to integration and humanization. It also creates flexibility and resilience in the ego. This is what constitutes the capacity to hold up against recollectivization, not to be swept away by the seductive forces of the collective into mass-mindedness.

By recollecting the voice of the ancestors calling upon us from underneath the layers of dust, and by listening to the call of the future, the future that always is a constantly newborn child, we can remain relatively integrated individuals even in the transiency of the present.

The human soul needs to be rediscovered, reenlivened after what Robert Sardello so aptly describes as having "withdrawn into the stillness that presides over the realm of death, as a consequence of the atomic bomb.[306] That is, faced with apocalypse and annihilation, the soul may freeze into silence. How do we thaw the soul and restore the ego's rootedness in the Self, in the world, in the world of Self and Soul?

This question has been raised by social philosophers, such as Hans Jonas[307], who elaborates on the need for ethics and responsibility in our technological age. Yet unfortunately, as Neil Postman has sadly pointed out, many psychotherapists have scant knowledge of the humanities, and for their field they are not required to have that knowledge."[308] Without it, what do they know of the many images of soul and human relationship that one encounters in reading literature, religion, philosophy, history, and even biology? Dullness and ignorance seem to have permeated psychotherapy and unfortunately many other fields, as well. Without the exposure to a wide breadth of humanity's knowledge, including its history and deep well of imagination, one cannot begin to contribute to the discussion about soul, its reaction to technology, or to the ethics and responsibility so sorely needed to contain the shadow of technological advancement.

I believe healing can take place by consciously looking into the shadow. The Hebrew word for shadow is *tzel*, צל, and the word for rescue is *hatzala*, הצלה; there is no rescue without courageously facing the shadow; this, in order to reenable the psyche's self-

---

[306] Sardello, 1999, 119ff.

[307] Hans Jonas, *The Imperative of Responsibility: In Search of an Ethics for the Technological Age.* Chicago, IL: University of Chicago Press, 1984.

[308] Neil Postman, 1993, p. 87.

regulation, creating a sense of relatedness and meaning. Through shadow recognition one's ego is strengthened, and projection onto the other is no longer psychologically necessary.

Furthermore, the path to healing consists in reconnecting and recognizing the archetypal dimension, archetypal fields and energies. This makes place and space sacred. And, finally, to tell the stories of depth and meaning, Eros, and reflection, a return to the deep well of memory is needed to embrace the Depth of Silence, the depth and the silence, from which, as the myth tells us, Wisdom may arise.

Let me emphasize *embrace*, that is, relatedness, because, as the clever woman Diotima from Mantinea tells Socrates, neither the wise nor the foolishsearch for wisdom, but *Eros* is a lover of wisdom, "lovers of wisdom being the intermediate class between the wise and the foolish."[309] As Jung says:

> There is a thinking in primordial images, in symbols which are older than the historical man, which are inborn in him from the earliest times, and, eternally living, outlasting all generations, still make up the groundwork of the human psyche. It is only possible to live the fullest life when we are in harmony with these symbols; wisdom is a return to them.[310]

By relating to and being touched by depth and silence, images can emerge from deep emotions, and, in turn, the images that rise from our internal depths can give shape to and express the complexity of emotions, such as the shades of erotic anxiety in an Ingmar Bergman movie.[311]

In the shallowness and the noise of dislocated transiency— not to be confused with profound movement and change—the images of wisdom do not have the freedom to emerge from the

---

[309] Plato, *Symposium and the Death of Socrates*, (Symposium 204), p. 38.
[310] Stages of Life, CW 8, par. 794.
[311] Cf. Shalit, Jung Journal, par. 2. P. 96.

depths of silence. I believe that it is by attending to the images of interiority that we sense the need to tend to the world in which the images of the psyche reside. Thus, perhaps, we become more able to induce our images with *meaning*, as generated by the archetype of the Self.

Rilke writes in his *Letters to a Young Poet*, "You must give birth to your images. They are the future waiting to be born."[312] Through our dreams, images, and imaginations we not only rely on thinking but we feel into and embody the future we want to create.

---

[312] Rainer Maria Rilke, *Letters to a Young Poet*, Harvard University Press, Cambridge, Mass: 2011.
https://www.goodreads.com/quotes/search?utf8=%E2%9C%93&q=You+must+give+birth+to+your+images.+They+are+the+future+waiting+to+be+born.&commit=Search

Bibliography

Adorno, T. (1973). *Negative dialectics*. (E.B. Ashton, Trans.). London: Routledge.

Anders, G. & Eatherly, C. (1961). *Burning conscience, The case of the Hiroshima pilot, Claude Eatherly, told in his letters to Gunther Anders*. New York, NY: Monthly Review Press.

Ronnberg, A. (Ed.). & Archive for Research in Archetypal Symbolism (2010). *The book of symbols: Reflections on archetypal images*. Los Angeles, CA: Taschen.

Arendt, H. (1970). *On violence*. New York, NY: Harcourt.

Arendt, H. (1994). *Eichmann in Jerusalem*. Harmondsworth, UK: Penguin.

Baudrillard, J. (1983). *Simulations*. New York: Semiotext.

Baudrillard, J. (2009). *Why hasn't everything disappeared?* London: Seagull books.

Bawaya, M. (2007). Saving the Mirador Basin. In *American Archaeology*, 11(3), 18-25.

Bicentennial Man, (1999). Screenplay by Nicholas Kazan; Directed by Chris Columbus. Based on the short story by Isaac Asimov, (1976). In *The Bicentennial man and other stories*, New York, NY: Doubleday.

Benjamin, W. (2002). The Storyteller, in *Selected Writings, vol.3*. In H. Eiland & M. W. Jennings (Eds.). (E Jebcott, H. Eiland. & Others, Trans.). Cambridge, MA: Harvard University Press.

Blake, W. (1965). *The complete poetry and prose of William Blake.* David Erdman (Ed.) New York, NY: Bantam Doubleday Dell.

Breuer, J. & Freud, S. (1991). *Studies on hysteria.* Harmondsworth, UK: Penguin.

Buber, M. (1969). *Leket: From the treasure house of Hassidism.* Jerusalem, Israel: WZO.

Buber, M. (1991). *Tales of the Hasidim.* New York, NY: Schocken Books.

Carter, L. (2008, March 17). Australian Auctions Entire Life on eBay. In *The Telegraph.* http://www/telegraph.co.uk/news/worldnews/australiaandthe pacific/1581952/Australian-auctions-entire-life-on-eBay.html.

Calvino, I. (1997). *Invisible cities.* London, England: Vintage.

Cheetham, T. (2003). *The world turned inside out: Henry Corbin and Islamic mysticism.* Woodstock, CT: Spring Journal Books.

Christenson, A. J. (2003). *Popol Vuh: Sacred book of the Maya.* New York, NY: O Books.

Descartes, R. (1880). *The method, meditations, and selections from the principles of Descartes.* Edinburgh, Scotland, and London, England: William Blackwood.

Deutsch, H. (1942). Some forms of emotional disturbance and their relationship to schizophrenia. In *Psychoanalytic Quarterly,* 11:301-321.

Dictionary of the history *of ideas*; http://onlinebooks.library.upenn.edu/webbin/book/lookupid?key=olbp31715

Dobson, A. (1883). *Old world idylls and other verses.* London, England: Kegan Paul, Trench, Trubner & Co.

Douthat, R. (2009, Oct. 4). Perpetual revelations: How ideas and practices from the past might guide believers in a modern approach to God. New York Times Book Review.

Dürrenmatt, F. (1964). *Four plays 1957-62.* London: Jonathan Cape.

Edinger, E. (1996). *The Aion lectures: Exploring the Self in C.G. Jung's Aion.* Toronto, Canada: Inner City Books.

Eliade, M. (1959). *The sacred and the profane: The nature of religion.* (Willard R. Trask, Trans.). New York, NY: Harcourt, Inc.

Eyes Wide Shut. (1999). Screenplay by Stanley Kubrick & Frederic Raphael; Directed by Stanley Kubrick, Based on the novel by Arthur Schnitzler, Dream Story, in *Eyes Wide Shut/Dream Story*. New York, NY: Warner Books.

Freud, S. (1957). On transience. In *On the history of the psycho-analytic movement: Papers on metapsychology and other works*. Vol. XIX, London, England: The Hogarth Press.

Freud, S. (1967). *On the history of the psycho-analytic movement, papers on metapsychology, and other works, SE 14.* New York, NY: W. W. Norton.

Freud, S. (1974). *Introductory lectures on psycho-analysis, SE 15.* (James Strachey, Trans.). London, England: Penguin Books.

Friedländer, S. (2013). *Franz Kafka, the poet of shame and guilt.* New Haven, CT, and London, England: Yale University Press.

Fromm, E. (1965). *The sane society.* New York, NY: Fawcett Premier.

Heraclitus. *The fragments of Heraclitus*, Fragment 121, p. 112. (G. T. W. Patrick, Trans.). Digireads.com Classic.

Huizinga, J. (1949). *Homo ludens: A study of the play-element in human culture.* London, England: Routledge & Kegan Paul.

Greene, N. A. (2008). *Horses at work: Harnessing power in industrial America.* Cambridge, MA: Harvard University Press.

Graves, R. (1992). *The Greek myths.* Harmondsworth, England: Penguin.

Hillman, J. (1992). *Re-Visioning psychology.* New York, NY: Harper Perennial.

Idel, M. (1990). *Golem: Jewish magical and mystical traditions: On the artificial anthropoid.* Albany, NY: State University of New York Press.

Idel, M. (1995). Sexual metaphors and praxis in the Kabbalah, In Mortimer Ostow, *Ultimate intimacy: The psychodynamics of Jewish Mysticism.* London, England: Karnac.

Jamison, L. (2017, Dec.). The Digital Ruins of a Forgotten Future, *The Atlantic.* Retrieved from https://www.theatlantic.com/magazine/archive/2017/12/second-life-leslie-jamison/544149/.

Jewish Study Bible, (2nd Ed). Jewish Publication Society.

Jonas, H. (1984). *The imperative of responsibility: In search of an ethics for the technological age.* Chicago, IL: University of Chicago Press.

Joseph, B. (1988). Projective identification: Clinical aspects, in Joseph Sandler (Ed.) *Projection, Identification, Projective Identification.* Oxford, England: Karnac.

Joy, B. (2000). Why the future doesn't need us, *Wired,* wired.com/wired/archive/8.04/joy

Jung, C.G. (1917). *Collected papers on analytical psychology.* New York, NY: Moffat Yard and Co.

Jung, C.G. (1953-1979). *The collected works.* Princeton, NJ: Princeton University Press.

Jung, C.G. Jung. (1965). *Memories, dreams, reflections.* New York:, NY Vintage Books.

Jung, C.G. (1976). *Psychological types.* Princeton, NJ: Princeton University Press.

Jung, C.G. (1980). *Archetypes of the collective unconscious.* Princeton, NJ: Princeton University Press.

Jung, C.G. (1989). *Psychology and religion: West and east.* (2nd Ed.) Princeton, NJ: Princeton University Press.

Jung, C.G. (2009). *The red book: Liber novus.* New York, NY; W.W. Norton Press.

Jung, C.G. 1977). *C.G. Jung speaking: Interviews and encounters.* William McGuire and R. F. C. Hull, (Eds.), Princeton, NJ: Princeton University Press.

Jung, C.G. & Neumann, E. (2015). *Analytical psychology in exile: The correspondence of C.G. Jung and Erich Neumann.* M. Liebscher, (Ed.). Princeton, NJ: Princeton University Press.

Kafka, F. (1991). *The blue octavo notebooks.* Max Brod, (Ed.). Cambridge, MA: Exact Change.

Kafka, F. (1971). The metamorphosis. In F. Kafka, *The complete stories.* New York, NY: Schocken.

Kaplan, A. (1997). *Sefer Yetzirah: The book of creation.* San Francisco, CA: Weiser Books.

Kass, L. & Wilson, J. (1998). *The ethics of human cloning.* Washington, DC: AEI Press.

Kelly, K. (2010). *What technology wants*. New York, NY: Viking, Penguin Group.

Kelly, K. & Johnson, S. (2010, October). Where ideas come from. *Wired*, 122-124.

Kurzweil, R. (2005). *The singularity is near: When humans transcend biology*. New York, NY: Viking.

Laplanche, J, & Pontalis, J. B. (1988). *The language of psycho-analysis*. London, England: Karnac Books.

New Larousse Encyclopedia of Mythology. (1987). Twickenham, England: Hamlyn.

Levi, P. (1987). *If this is a man*. London: Abacus.

Maimonides, M. (1963). *The guide for the perplexed.* (S. Pines, Trans.). Chicago: The University of Chicago Press.

Masson, James. (1990). *Against Therapy*. London, England: Fontana.

Maier, C. A. (2012). *Healing dream and ritual: Ancient incubation and modern psychotherapy*. Einsiedeln, Switzerland: Daimon Verlag.

Markoff, J. (2013, May 21). In 1949, He imagined an age of robots. *New York Times*.

Mayor, T. (2011, Jan. 30). *Computerworld*. Retrieved January 30, 2011.
Asperger's and IT: Dark secret or open secret? From:http://www.computerworld.com/s/article/9072119/Asperger_s_and_IT_Dark_secret_or_open_secret_?taxonomyId=14&pageNumber=2

Meyrink, G. (1995). *The Golem*. Cambs, England: Dedalus.

Mumford, L. (1967). *Technics and civilization: The myth of the machine, Volume One.* New York, NY: Harcourt Brace Jovanovich.

Mumford, L. (1986). *The Lewis Mumford reader*. Donald L. Miller (Ed). The Foundations of Eutopia. New York, NY: Pantheon Books.

Mumford, L. (1922). *The Story of utopias*. NY: Boni and Liveright.

Munroe, R. *http://what-if.xkcd.com/23/*.

Neumann, E. (1989). *The place of creation*. Princeton, NJ: Princeton University Press.

Neumann, E. (1970). *The origins and history of consciousness.* Princeton, NJ: Princeton University Press.

Neumann, E. (2016). *Jacob and Esau: On the collective symbolism of the brother motif.* Erel Shalit. (Ed). Asheville, NC: Chiron.

Niederland, W. (1968). Clinical observations on survivor syndrome. In *International Journal of Psychoanalysis*, 49: 313-315.

Nietzsche, F. (2001). *The gay science.* B. Williams (Ed). Cambridge MA: Cambridge University Press.

Onians, R. B. (1951). *The origins of european thought.* Cambridge, MA: Cambridge University Press.

Ovid. (1986). *Metamorphoses.* (A. D. Melville, Trans.). Oxford, England: Oxford University Press.

Pirandello, L. (1998). *Six characters in search of an author.* New York, NY: Signet Classic, Penguin Books.

Plato. (1977). *Timaeus and critias.* (D. Lee. Trans.) Harmondsworth, England: Penguin.

Plato. (1997). *Symposium and the death of Socrates.* (T Griffith, Trans.). Hertfordshire, England: Wordsworth.

Postman, N. (1993). *Technopoly: The surrender of culture to technology.* New York, NY: Vintage.

Rappoport, A. (1995). *Ancient Israel: Myths and legends.* London, England: Senate.

Rig Veda, (2005). Reprinted Ed. Edition. New York, NY: Penguin Classics.

Rilke, R. M. (2010). Testament in *the inner sky: Poems, notes, dreams.* (Daimon Searls, Trans.). Boston, MA: David R. Godine.

Ronnberg, A. (Ed.). (2010). *The book of symbols.* Germany: Taschen.

Sadeh, P. (1989). *Jewish folktales.* (The Hebrew by Hillel Halkin, Trans.). New York, NY: Doubleday.

Sarna, N. (1970). *Understanding Genesis: Through Rabbinic tradition and modern scholarship.* New York, NY: Schocken.

Sartre, J. P. (1989). *No exit and three other plays.* (Stuart Gilbert, Trans.). New York, NY: Vintage International Books.

Schele, L., & Matthews, P, (1998). *The code of kings: The language of seven sacred Maya temples and tombs.* New York, NY: Simon & Schuster.

Scholem, G. (1969). *On the Kabbalah and its symbolism,* New York, NY: Schocken.

Segal, D. (2013, June 2). This man is not a cyborg. Yet. *The New York Times.*

Shalit, E. (2002). *The complex: Path of transformation from archetype to ego.* Toronto, Canada: Inner City Books.

Shalit, E. (2004). *The hero and his shadow: Psychopolitical aspects of myth and reality in Israel.* Dallas, TX: University Press of America.

Shalit, E. (2008). *Enemy, cripple, beggar: Shadows in the hero's path.* Hanford, CA: Fisher King Press.

Shalit, E. (2010). *Self, meaning and the transient personality.* XVlllth.

Shalit, E. (2010, August). International Congress for Analytical Psychology.

Shalit, E (2010). Destruction of the image and the worship of transience. In *The Jung Journal* 4(1), San Francisco, CA: Allen Press.

Shalit, E. (2011a). *The cycle of life: Themes and tales of the journey.* Hanford CA: Fisher King Press.

Shalit, E. (2011b). *The hero and his shadow: Psychopolitical aspects of myth and reality in Israel.* Hanford, CA: Fisher King Press.

Sherwood, V, R. & Cohen, C. P. (1994). *Psychotherapy of the quiet borderline patient: The as-if personality revisited.* London, England: Jason Aronson.

Sontag, S. (2003). *Regarding the pain of others.* New York, NY: Picador.

Sontag, S. (2001). *On photography.* New York, NY: Picador.

Sontag, S. (2007). *At the same time: Essays and speeches.* Paolo Dilonardo and Anne Jump, (Eds.). Foreword by David Rieff. New York, NY: Farrar Strauss Giroux.

Snyder, T. (2017). *On tyranny: Twenty lessons from the twentieth century.* New York, NY : Tim Duggan Books.

Stein, R. (1974). *Incest and human love: The betrayal of the soul in psychotherapy*. Baltimore, MD: Penguin.

Stevens, A. (1995). *Private myths: Dreams and dreaming*. Cambridge, MA: Harvard University Press.

Taleb, N, (2010). *The black swan*. New York, NY: Random House.

Tedlock, D. (1985). *Popol Vuh: The definitive edition of the Mayan book of the dawn of life and the glories of gods and kings*. New York, NY: Touchstone Books.

Tripp, E. (1970). *Handbook of classical mythology*. New York, NY: Meridian.

The Arabian Nights. (2001). New York, NY: The Modern Library.

Ullrich, V. (2016). *Hitler: Ascent 1889-1939*. New York, NY: Knopf.

Valéry, P. (1989). *Degas, Manet, Morisot*. (David Paul, Trans.). Princeton, NJ: Princeton University Press.

Vilnay, Z. (1973). *Legends of Jerusalem*. Philadelphia, PA: Jewish Publication Society of America.

Von Franz, M. L. (1964). The process of individuation, pp. 158-229. In C.G. Jung, M. L. von Franz, Joseph L. Henderson, Jolande Jacobi, Aniela Jaffé, *Man and his Symbols*. New York, NY: Doubleday.

Von Franz, M. L. (1995). *Creation myths*. Boston, MA: Shambhala Press.

Walker, B. (1983). *The woman'encyclopedia of myths and secrets*. San Francisco, CA: Harper Collins.

Wiener, N. (1964). *God & golem, inc.* Cambridge, MA: MIT Press.

Wiesel, E. (1982). *Legends of our time*. New York, NY: Schocken Books.

Zola, E. (1972). *L'Assommoir*. (Leonard Tancock, Trans.). NY: Harmondsworth, Penguin.

Zola, E. (1998). *Nana*. (Douglas Parmee, Trans.). Oxford: Oxford University Press.

Index

Contributing Author - Nancy Swift Furlotti

Nancy Swift Furlotti, Ph.D. is a Jungian Analyst in Aspen, CO. She is a past President of the C.G. Jung Institute of Los Angeles and founding member of the C.G. Jung Institute of Colorado. She is also a member of the Inter-Regional Society of Jungian Analysts, and teaches and lectures in the US and internationally. Her articles "The Archetypal drama in Puccini's *Madam Butterfly*" and "Tracing a Red Thread: Synchronicity and Jung's Red Book" have recently been published in *Psychological Perspectives*. She also has a chapter, "Angels and Idols: Los Angeles, A City of Contrasts" in Tom Singer's (ed.) book, *Psyche and the City: A Soul's Guide to the Modern Metropolis.* She edited a book with Dr. Erel Shalit entitled, *The Dream and its Amplification.* She has one chapter in the book, *Turbulent Times, Creative Minds: Erich Neumann and C.G. Jung in Relationship,* and two chapters in *A Clear and Present Danger: Narcissism in the Era of President Trump,* Narcissism in the Home, and Narcissism in our Collective Home, American Culture. She contributed a chapter called, Encounters with the Animal Soul: A Voice of Hope for Our Precarious World, in the book, *Jung's Red Book for Our Time: Searching for Soul under Postmodern Conditions,* edited by Murray Stein and Thomas Arzt.

Dr. Swift Furlotti has a deep interest in exploring the manifestations of the psyche through dreams and myths, with a specific focus on the dark emanations from the psyche. A longstanding focus of research is on Mesoamerican mythology. Her dissertation was titled, *A Jungian Psychological Amplification of the Popol Vuh, the Quiché Maya Creation Myth.* Her interest in exploring symbols and deepening her understanding of Jung, have landed her on two foundations: The Philemon Foundation, where she is a founding board member and served as co-President, and ARAS (Archive for Research in Archetypal Symbolism). She is on a board member of the Kairos Film Foundation that oversees the Remembering Jung Video Series, 30 interviews with Jungian analysts, and the films, *A Matter of Heart* and *The World Within,* and continues to disseminate Jungian ideas through film. Dr. Swift Furlotti established the Carl Jung Professorial Endowment in Analytical Psychology at the Semel Institute for Neuroscience and Human Behavior at UCLA. She is also a board member of the Foundation for Anthropological Research & Environmental Studies (FARES) and is also a member of the board of Pacifica Graduate Institute and a member of the Mercurius Film Prize Committee.

CPSIA information can be obtained
at www.ICGtesting.com
Printed in the USA
FFHW020808140219
50506566-55775FF